Militarizing Artificial Intelligence

This book examines the military characteristics and potential of Artificial Intelligence (AI) in the new global revolution in military affairs.

Offering an original perspective on the utilization, imagination, and politics of AI in the context of military development and weapons regulation, the work provides a comprehensive response to the question of how we might reflect on the AI revolution in warfare and what can be said about the ways in which this has been handled. In the first part of the book, AI is accommodated, both theoretically and empirically, in the strategic context of the 'Revolution in Military Affairs' (RMA). The book offers a novel understanding of autonomous weapons as multi-layered composite systems, pointing to a complex, non-linear interplay between evolutionary and revolutionary dynamics. In the second section, the book provides an impartial analysis of the related politics and operations of power, whereby increases in military budgets and R&D of the great powers are met and countered by advocacy networks and scientists campaigning for a ban on lethal autonomous weapons. As such, it moves beyond popular caricatures of 'killer robots' and points out some of the problems which result from over-reliance on such imagery.

This book will be of much interest to students of strategic studies, critical security studies, arms control and disarmament, science and technology studies, and general International Relations.

Nik Hynek is a professor specializing in security studies at the Department of Security Studies, Faculty of Social Sciences at Charles University. He leads the inter-scientific Charles University Research Centre of Excellence dedicated to the topic of 'Human-Machine Nexus and the Implications for the International Order'.

Anzhelika Solovyeva is a lecturer specializing in strategic studies at the Department of Security Studies, Faculty of Social Sciences at Charles University. Her latest monograph, co-authored with Nik Hynek, is *The Logic of Humanitarian Arms Control and Disarmament* (2020).

Routledge Studies in Conflict, Security and Technology

Series Editors: Mark Lacy, Lancaster University, Dan Prince, Lancaster University, and Sean Lawson, University of Utah

The *Routledge Studies in Conflict, Technology and Security* series aims to publish challenging studies that map the terrain of technology and security from a range of disciplinary perspectives, offering critical perspectives on the issues that concern publics, business and policymakers in a time of rapid and disruptive technological change.

National Cyber Emergencies
The Return to Civil Defence
Edited by Greg Austin

Information Warfare in the Age of Cyber Conflict
Edited by Christopher Whyte, A. Trevor Thrall, and Brian M. Mazanec

Emerging Security Technologies and EU Governance
Actors, Practices and Processes
Edited by Antonio Calcara, Raluca Csernatoni and Chantal Lavallée

Cyber-Security Education
Principles and Policies
Edited by Greg Austin

Emerging Technologies and International Security
Machines, the State and War
Edited by Reuben Steff, Joe Burton and Simona R. Soare

Militarizing Artificial Intelligence
Theory, Technology and Regulation
Nik Hynek and Anzhelika Solovyeva

For more information about this series, please visit: https://www.routledge.com/Routledge-Studies-in-Conflict-Security-and-Technology/book-series/CST

Militarizing Artificial Intelligence

Theory, Technology, and Regulation

Nik Hynek and Anzhelika Solovyeva

Routledge
Taylor & Francis Group
LONDON AND NEW YORK

First published 2022
by Routledge
4 Park Square, Milton Park, Abingdon, Oxon OX14 4RN

and by Routledge
605 Third Avenue, New York, NY 10158

Routledge is an imprint of the Taylor & Francis Group, an informa business

British Library Cataloguing-in-Publication Data
A catalogue record for this book is available from the British Library

Library of Congress Cataloging-in-Publication Data
Names: Hynek, Nik, author. | Solovyeva, Anzhelika, 1994– author.
Title: Militarising artificial intelligence : theory, technology and regulation / Nik Hynek and Anzhelika Solovyeva.
Other titles: Militarizing artificial intelligence
Description: Abingdon, Oxon ; New York : Routledge, [2022] |
Series: Routledge studies in conflict, security and technology |
Includes bibliographical references and index.
Identifiers: LCCN 2022000868 (print) | LCCN 2022000869 (ebook) |
ISBN 9780367492854 (hardback) | ISBN 9780367492878 (paperback) |
ISBN 9781003045489 (ebook)
Subjects: LCSH: Artificial intelligence—Military applications. |
Artificial intelligence—Military applications—Government policy. |
Weapons systems—Government policy. | Autonomous weapons systems (International law)
Classification: LCC UG479 .H96 2022 (print) | LCC UG479 (ebook) |
DDC 359.00285/63—dc23/eng/20220307
LC record available at https://lccn.loc.gov/2022000868
LC ebook record available at https://lccn.loc.gov/2022000869

ISBN: 978-0-367-49285-4 (hbk)
ISBN: 978-0-367-49287-8 (pbk)
ISBN: 978-1-003-04548-9 (ebk)

DOI: 10.4324/9781003045489

Typeset in Goudy
by codeMantra

Contents

About the Authors

Nik Hynek is a professor specializing in security studies at the Department of Security Studies, Faculty of Social Sciences at Charles University. He received his research doctorate from the University of Bradford and was previously affiliated with SIWPS at Columbia University, LSE, ANU, Carleton University, and Ritsumeikan University. He leads the inter-scientific Charles University Research Centre of Excellence dedicated to the topic of 'Human-Machine Nexus and the Implications for the International Order'.

Anzhelika Solovyeva is a lecturer specializing in strategic studies at the Department of Security Studies, Faculty of Social Sciences at Charles University. She also works as a doctoral researcher at the Charles University Research Centre of Excellence dedicated to the topic of 'Human-Machine Nexus and the Implications for the International Order'. Her latest monograph, co-authored with Nik Hynek, is *The Logic of Humanitarian Arms Control and Disarmament* (2020).

Acknowledgements

The preparation of this manuscript was undertaken with financial support from Charles University, UNCE 'Human-Machine Nexus and the Implications for the International Order' (UNCE/HUM/037). The book would have been poorer without the comments of two anonymous readers whose challenges and insights we gratefully acknowledge too. We owe a special debt to Xing Su for kindly proofreading the entire manuscript. We would also like to acknowledge that Chapter 4 is an updated and substantially rewritten version of Nik Hynek and Anzhelika Solovyeva's 'Going Beyond the "Killer Robots" Debate: Six Dilemmas Autonomous Weapon Systems Raise', *Central European Journal of International and Security Studies* 12(3) (2018): 166–208; and that an earlier version of Chapter 6 was previously published as Nik Hynek and Anzhelika Solovyeva's 'Operations of Power in Autonomous Weapon Systems: Ethical Conditions and Socio-Political prospects', *AI & Society: Knowledge, Culture and Communication* 36 (2021): 79–99. Finally, we wish to thank our families for their long-standing support and encouragement of our academic endeavours.

Introduction

In recent years there has been a breathtaking increase in discussions about the militarization of artificial intelligence (AI). The development of militarized AI is now seen as the next frontier for technological innovation, great power competition, and political activism. This book offers an original perspective on the utilization, imagination, and politics of AI in the context of military development and weapons regulation. Its main goal is to provide a comprehensive answer to the compelling, yet contentious, problem of *how we can reflect on the current revolution in military affairs (RMA), the so called 'AI-RMA', and what can be said about the ways in which it has been handled*. RMA theory, securitization theory, and regime theory provide a robust theoretical basis from which to analyse the transformative potential of AI. The motivation is simple: as evident from wide-ranging debates, as well as active and sustained regulatory efforts, the future of warfare is clearly linked to AI. Over the past decade increasing attention has been given to 'killer robots', defined as the emerging category of artificially intelligent weapons which will be autonomous to the extent of being able to decide for themselves when and whom to kill. Transnational policy advocacy has become very active, proliferating a particular imagery of the weapons, with science being used for strategic manipulation and legitimizing prohibitory agendas. There is a powerful sense of urgency in dealing with the new, potentially perilous challenges posed by militarized AI. Still more importantly, there is a growing demand for impartial judgment and a balanced assessment of the risks.

It is for these reasons that we recreate the existing and emerging political construction of 'killer robots' or, less colloquially, autonomous weapons systems (AWS). In doing so, we start anew, moving beyond popular caricatures of 'killer robots' and even rejecting the term itself, and point out some of the problems which result from over-reliance on the wrong imagery. We offer a distinct theoretical framework that allows us to fit the AI-RMA and AWS into a historically anchored, evolutionary, and multi-layered context. What results is a brand new conceptualization of artificially intelligent weapons, treated herein as *composite systems*. Through these lenses, we analyse and illustrate how AI is being incorporated into weapons systems, both existing and under development, in China, Russia, and the US. The analysis has important implications for understanding, inter alia, what it means to be a great power in the twenty-first century. Finally, as

DOI: 10.4324/9781003045489-1

its major contribution, this book embeds these emerging, revolutionary technologies in a wider terrain where the established global politico-economic structures encounter humanitarian action and ethical discourses.

The book is divided into two parts. The general flow of the argument from Part I to Part II demonstrates how the analysis becomes gradually narrower: from the fairly broad AI-RMA to dilemmas and regime dynamics related directly to AWS. Nevertheless, there is a single line of argument that underlies both parts and the respective chapters. The following questions are systematically answered: How can we reflect on the transformative potential of AI through the lens of RMA? How can AI contribute to our understanding of RMA? How has the AI-RMA manifested itself in the real world, and particularly in the context of great power competition between China, Russia, and the US? What has been the most actively pursued and discussed regulatory agenda? What are the pitfalls involved in the discursive and visual constructions of the 'threat' of weaponized AI, which supposedly underlies AWS? What counter-arguments have been put forward? Why has there been no ban and what are the prospects of a prohibitory, or even regulatory, regime in this issue area? The central problem this book poses can only be addressed after these questions are answered.

The major contribution of this book lies in presenting a theoretically informed, empirically rich, multi-dimensional, and normatively impartial analysis of the militarization of AI. The greatest contribution to the empirical literature consists in bridging the existing knowledge about AI and AWS with theoretically oriented literature on securitization, international security regimes, and RMA. Only by doing so can we systematically and comprehensively address the ongoing processes of change at all relevant levels: in military affairs and broader society, in technology and organization, in discourse and practice, in the perception of reality and actual reality, and in social and legal norms. No less important is the detailed and nuanced analysis of the relationship between AI and AWS. In particular, we discuss and illustrate the difference between *militarized* AI, AI-*assisted* weapons, and yet non-existent *fully autonomous lethal* AI. While all these can dramatically increase targeting efficiency and add to the autonomous capacity of weapons systems, only the last category is directly associated with the envisioned – so compelling and so frightening – purpose of AWS. It is the main subject of discussion in Chapters 3 and 4. What can be seen as yet another essential contribution is that the book presents a balanced mix of political insight and impartial, fact-driven, emotionless analysis of AWS. As efforts to establish serious negotiations for a complete ban on AWS and acts of resistance have intensified, the impartiality of experts has often been compromised by their direct engagement in policy making and political manipulation. We deal with this problem in Chapters 4 and 5. Our principled position is that we do not take political sides in the debate on AWS. We uphold the highest epistemic standards, which is the best strategy for not falling into bias traps. Simply put, we are committed to analysing things 'from the outside', i.e., from a purely scientific, and not political, perspective. Our critical review of certain policy choices throughout this book is, therefore, not meant to express our normative or political attitude but rather to contribute to a better social scientific understanding of the phenomena examined.

The book also identifies considerable heuristic limitations and blind spots with respect to the application of RMA theory, securitization theory, and regime theory to AI and AWS. Its manifold contributions to theoretically oriented literature are detailed below. The first fundamental finding, and simultaneously our starting point, is that there is no general theory of RMA. RMA theory rather consists of different, often disjointed and contradictory, theoretical approaches, as shown in Chapter 1. We borrow much of the intellectual framework from organization theory and new institutionalism to integrate these clearly fragmentary approaches in a simple, flexible, and comprehensive way. What we offer as a result is a truly integrated approach to RMA. Chapter 2 inquires into the technological parameter of RMA in light of AI. While thinking *through*, and not just about, the most recent technological innovations, we draw attention to the fact that several generations of technology mesh with one another and coalesce in what we call *composite systems*. The proposed conceptual model firmly incorporates and accurately grasps the transformative potential of AI. However, it will be of much wider interest and applicability. Its main contribution is in bridging the gap between existing typical, ideal representations of military-technological innovation ('generations', 'ages', etc.) and the messy reality. Last but certainly not least, the model avoids technological determinism and embeds technological change within a wider social and organizational context. In doing so, it captures and conceptualizes the link between technological and military change. Overall, the book offers an innovative conceptualization bringing together organizational, technological, and historical perspectives to RMA.

In addition, the book offers a distinct and unique contribution to securitization theory. Chapter 5 forges an original way of thinking about the process and existing tools of securitization. Primarily, it explores and conceptualizes the dynamics of *over*-securitization, an area that unfortunately remains underexplored. Over-securitization, as defined in this chapter, is about securitization efforts that begin as useful and gradually become counter-productive. Two analytical tools are developed and applied to understand how over-securitization plays out in everyday practice: the concepts of *hybridization* and *grafting*, as explained in detail below. The arguments and findings presented in Chapter 5 are meant to contribute to a deeper and more nuanced understanding of the *continuum* of securitization. Therefore, we further problematize the classic dichotomy that securitization moves can *either* succeed *or* fail.

The book makes a strong contribution to regime theory as well. The *power-analytical approach*, originally developed by one of the authors elsewhere and further sophisticated in Chapter 6, cuts through the three waves of regime theorization using the conceptualization of power. Making the concept of power central, we at least partially compensate for the limits experienced by each of the three waves and find ourselves in a heuristically favourable position. Although not aiming for theory synthesis, we are able to investigate how different types of forces, discourses, and interests, all related to interactions among actors and individually pertinent to different theories, cross-fertilize and synergize. At the same time, the available knowledge on *power* is enriched with a new interpretation of synergies between different types of power: productive, structural, compulsory, and institutional. What is more, we offer a nuanced graphical representation of their complex interactions in practice. Generally speaking, the book offers new insights, new explanations, and new understandings of empirical data, treated in an original and stimulating manner.

We now turn to a more detailed elaboration of the structure of this book. Part I disaggregates the AI-RMA along multiple dimensions: social, technological, historical, and spatial. It consists of three consecutive chapters. **Chapter 1** begins by accommodating AI into the conceptual and analytical framework of RMA. In treating the existing framework systematically, this chapter clearly demonstrates that it comprises sporadic, piecemeal, compartmentalized, and sometimes even contradictory approaches. The second part of the chapter proposes an alternative, more inclusive approach to identifying and analysing the key components and processes of what can eventually be defined as RMA. Inter alia, it offers a more flexible and more integrated way of thinking about military innovations, both evolutionary and revolutionary, and broader socio-political and economic dynamics. Organization theory and institutional logics are utilized to integrate heretofore disjointed efforts to theorize RMA. The primary purpose of introducing a new, refined framework is to bridge the gap between the theorized reality and the actual reality of the AI-RMA. However, we argue for its applicability in a wider range of cases and believe this novel perspective will add breadth and depth to the current understanding of RMA.

Chapter 2 identifies a range of flaws and gaps in the existing body of knowledge about revolutionary technologies themselves. In doing so, it reverses the main question posed in Chapter 1. The focus is not on whether or how AI *fits in with* the current conceptual framework of RMA. Instead, it is on whether and how AI *can itself contribute to* the existing literature and theories associated with RMA. This chapter calls attention more particularly to the technological parameter of RMA. AI reveals certain trends in technological development, as we show. These are taken as the basis to develop an integrated technological model which features four *layers* of systems capabilities in five warfighting domains: land, sea, air, outer space, and cyberspace. We argue and vividly demonstrate in the chapter that the long history of technological change can be best conveyed by a circular and layered representation. Hence, the proposed model goes beyond the existing, often linear, representations of technological change. It is designed to analyse what we call *composite systems*, a term we introduce in searching for a more essential conceptual model. This model not only captures the nature of technological change, it also conceptualizes and visualizes the *link* between technologies and RMA. No less important is the fact that it brings to light evolutionary characteristics of the AI-RMA, something that has often been suppressed in favour of the argument for a new military 'revolution'.

Chapter 3 uses the sets of criteria and theoretical assumptions developed in the previous two chapters to inquire into today's realities of the AI-RMA in China, Russia, and the US. Both evolutionary and revolutionary signs of the ongoing process of transformation are put under the microscope. The analysis is carried out on two different levels: the level of technologies and the level of military organizations. In regards to the former, this chapter shows how *composite systems* – the concept taken from a *diachronic* analysis of technological change in Chapter 2 – can also be analysed comparatively, that is *synchronically*. Technical characteristics of 12 composite systems with current and potential military applications are examined. Most importantly, perhaps, the analysis presented here clearly distinguishes between merely *militarized* AI, AI-*assisted* weapons, and yet non-existent *fully autonomous lethal* AI. With respect to the other level of analysis, the chapter also draws attention to the socio-political and institutional side of the AI-RMA. The focus is shifted to pervasive change in strategic concepts and doctrines, in the distribution of roles and responsibilities among military and closely related institutions, in training and operational routines, and the implications thereof. Case selection is driven by practical concerns. China, Russia, and the US are all great powers with their own competing visions of the world order. All of them manifest global aspirations predicated on geopolitics and geostrategy. The same three are the world's top military spenders and strategic competitors. Last but not least, each has openly declared its aspiration for global leadership in the field of AI.

Part II digs deep into the regime dynamics emerging around the AI-RMA. One of the key themes is the close connection, yet substantial difference, between AI and AWS, colloquially called 'killer robots'. **Chapter 4** still more meticulously elaborates on the difference between increasingly autonomous functions – including six *intelligent* functions – in weapons systems and as-yet imagined *fully autonomous lethal* AI. It discusses a range of terms by which increasingly autonomous weapons are defined and identifies 'killer robots' as the most pronounced, yet most problematic. Still more precisely, this chapter begins by investigating technical aspects of increasingly autonomous weapons, deployed and under development, and broadly discussing the gap between the real and the imagined. The most striking finding is that 'killer robots', often cited by campaigners as the emerging category of weapons that should be banned, only exist in the non-tangible, imaginary, and speculative realm, at least for the moment. The chapter then proceeds to present seven *dilemmas* in relation to fully autonomous weapons that need to be reconciled on the way towards some sort of regulatory response. In weighing the potential pros and cons of these weapons, it clearly shows that a simple outright ban, without such a discussion preceding it, will miss the complexity of the issue.

Chapter 5 focuses specifically on the Campaign to Stop Killer Robots. Its goal is to highlight and systematically analyse the apparent paradoxical effects of what we call *over*-securitization. The chapter shows that the Campaign's lack of wider political success stems principally, and rather counter-intuitively, from its highly efficient, yet excessive and eventually counter-productive, securitization strategies. Two concepts are introduced to define two different but interrelated mechanisms through which over-securitization operates: *hybridization* and *grafting*. The former helps demonstrate how complex hybridization of securitizing actors and target audiences produces circulatory, transepistemic, and post-truth representations of security. Therefore, the first and most obvious paradox concerns the relative diffusion of authority. The other concept serves to capture how a new security agenda can be reinforced through references to other security issues of immediate importance and even science fiction imagery. This brings us to the second paradox, which we illuminate by bringing attention to reductionist traps, overlapping security games, unfortunate parallels, and unproductive fictions. The underlying theme that is strongly foregrounded in this chapter is the tension between the *efficiency* and *effectiveness* of political advocacy with respect to weapons regulation.

Having discussed a number of problems with the very notion of 'killer robots' and the approach taken by the Campaign, the last chapter takes the argument one level further. **Chapter 6** scrutinizes the process by which the initiative on banning AWS enters the politically volatile terrain of arms control and disarmament. The major focus is on efforts, so far largely unsuccessful, to construct a new global regulatory order and reverse the tendency towards increasing autonomy in weapons. The concept of *power* provides a new way to understand the dynamics underlying the formation of international security regimes. The *power-analytical approach* utilizing four types of power – productive, structural, compulsory, and institutional – allows for a wide range of factors to be incorporated and theorized accordingly. In theoretical terms, the chapter explores how different forms of power interrelate and cross-pollinate in the workings of two social structures: ethics and law. From a practical point of view, it shows how prohibition-oriented policy advocacy meets great power politics. It makes particularly clear that the Campaign meets the blocking coalition on many levels: their discourses, structural relations, coercive strategies, and positions within relevant security-related institutions are all analysed in a comparative manner. The chapter then reflects on the prospects and limitations, as well as the ethical and legal intensity of the emerging regulatory framework in the issue area of AWS.

Part I

Artificial Intelligence and Dynamics of Military Transformation

1 Artificial Intelligence and the Revolution in Military Affairs

Introduction

We are told that we stand on the cusp of another revolution in military affairs (RMA). It is being brought about by a rapidly evolving, transformative, and potentially disruptive technology commonly known as artificial intelligence (AI). Dreams of human-like machines and the earliest documented examples of automata date back far into the past, but the first real efforts to build intelligent machines originated in the mid-twentieth century (Nilsson 2010: 71; Cave and Dihal 2018). It is often ignored, however, that there is no such a thing as AI in a complete form. This is, in fact, an umbrella term for a broad range of advanced technologies which allow the mimicry of human abilities in machines and breathe life into *what we know as* AI. Such technologies include but are not limited to machine learning techniques, artificial neural networks, computer vision, multisensor data fusion, as well as algorithms for natural language processing, pattern and speech recognition, knowledge representation, and reasoning (Nilsson 2010). The Internet of Things and Big Data facilitate progress towards AI (Hallaq et al. 2017). AI technologies underlie, among other things, the *intelligentization* of weapons, battlefield networks, intelligence, surveillance and reconnaissance capabilities, military logistics, and war games. This is where our focus lies and what this book analyses under the rubric of the AI-RMA. Since the concept of RMA is often evoked in relation to AI, this book begins with an analysis of whether and how AI *fits in with* what is conceptualized as RMA. In particular, we seek to understand if the existing intellectual framework can be applied to reflect on the transformative potential of AI. Another goal emerges as the analysis proceeds: to bring the existing theorization of RMA closer to reality.

RMA analytical framework: Accommodating AI

This section reviews the existing literature and unravels the phenomenon of RMA. The focus is on the readiness of its conceptual and historical dimensions to accommodate and explain the revolutionary potential of AI. It is fair to begin this discussion by noting that there has been no consensus on the definition of RMA. However, and here we part with Shimko (2010: 14), differences are often

DOI: 10.4324/9781003045489-3

a matter of degree. Though making the application of the concept itself less straightforward, they do not undermine its analytical significance and explanatory potential. What follows is a systematic inquiry into the various existing theories of RMA. In particular, we scrutinize and report on progress towards an integrated analytical framework. The following aspects are considered: the role of technology in generating military change, the nature of revolutionary outcomes, and the respective military history.

Components of change

It is broadly accepted in the existing literature that technology plays an important role in facilitating military revolutions. Yet all parties concur that it is not the only driving force (van Creveld 1991: 32; Toffler and Toffler 1993: 31–34; Rogers 2000: 30; Murray and Knox 2001: 12; Cohen 2004: 399; Adamsky 2010: 1). However, theorists differ in views as to the degree of technological impact. Some scholars treated technology as the *necessary condition* (Krepinevich 1994). They argued that it sets the *parameters of the possible* (Boot 2006: 43). Others were less resolute in their views. They assumed that technology *usually* (e.g., Fitzsimonds and van Tol 1994: 25) *normally* (e.g., Morgan 2000: 134) or *often* (e.g., Sloan 2002: 25) drives change in military art. Still more sceptical writers note it only *sometimes* and *not always* does so (Hundley 1999: 14; Horowitz 2010: 22). In other works of this kind, technology was conceived of as a *relatively insignificant* factor (Murray 1997). They believed it *rarely* plays a decisive role (Murray and Knox 2001: 180). Two approaches stand out as the leading sources of theorization. They differ with respect to the role of technology and assign different weights to different factors in the overall composition of change.

One body of literature takes technology seriously and theorizes the *requisites* of change. In spite of differences in formulation, their proposed set of indicators basically includes technological change, operational innovation, and organizational adaptation (Fitzsimonds and van Tol 1994: 25–29; Krepinevich 1994; Hundley 1999: 14–22 and 33; Morgan 2000: 135–138; Gilli and Gilli 2016: 56–60). It is well-tailored to analyse technology-driven revolutions, including both *military-technical revolutions* (MTRs) and *revolutions in military affairs*. The former is often traced back to the writings of Soviet military thinkers in the 1970s and 1980s (Krepinevich 2002: 1; Thompson 2011: 85); the latter is a more recent term and differs in the increasing recognition of doctrinal and organizational factors over the primacy of technology (Fitzsimonds and van Tol 1994: 25–26; Hundley 1999: 14–15; Morgan 2000: 134–135; Sloan 2002: 24–25). Adamsky (2010: 5) went even further to argue that there is also strategic culture, which provides the decisive context for one's RMAs. The logic of MTRs, though shelved for the present, sometimes manifests itself in different and more neutral terms such as *military revolutions* (Krepinevich 1994) or *military innovations* (Gilli and Gilli 2016: 82).

The other body of literature theorizes *potential* sets of indicators to get closer to reality. For them, technology is only marginally significant. Although it helps them to better capture the variety of military revolutions, there has been no agreement on the nature and composition of factors in the equation. For instance, Rogers (2000: 22–24) analysed *revolutions in military affairs* as changes in how war is fought, including technical, tactical, and strategic innovations. However, he argued that such changes in certain, but by no means all, cases have extremely wide-ranging social, economic, and political implications. Whenever it is the case, he used another term: *(full) military revolutions*. Knox and Murray (2001: 6–12) focused on *(great) military revolutions* recasting the entirety of society, the state, and the system of military organization. However, they acknowledged the existence of less profound transformations that either accompanied or followed such revolutions. They studied them as *revolutions in military affairs* and as a complex mix of tactical, organizational, doctrinal, and technological innovations. These two approaches are similar in terms of viewing military revolutions as much broader structural changes compared to RMAs, but they apparently differ in how they theorize the relationship between these two terms and phenomena. There are many more approaches in fact. For example, Toffler and Toffler (1993: 31–34) inquired into the transformation of society under the rubrics of *civilizational change*. Such large-scale transformations, in their opinion, entail *(true) military revolutions*. The latter involve profound changes in armed forces at various levels from technology, doctrine, and organization to tactics, strategy, logistics, and culture. Their approach did not override the possibility of *sub-revolutions*, potentially limited to certain contexts but also part of the total 'game'. Boot (2006: 26–28) drew attention to paradigm shifts associated with transformational technologies. He presented them as *ages*. Only when opportunities for new competitive advantages are exploited by societies and their armies do we experience *military revolutions*. The author associated them, inter alia, with changes in organization, strategies, tactics, leadership, training, and morale. Gray (2006: 16–28) distinguished six different 'contexts' shaping the prospects of occurrence and the character of what he called *revolutionary change in warfare*. These are strategic considerations, socio-cultural trends, economic conditions, technological capabilities, politics, and geography. Horowitz (2010: 23–42) analysed *military innovations* and conceptualized them as the product of one's choices and possibilities. Strategic choices, in his view, arise from geostrategic dynamics, international norms, domestic politics, and cultural openness. Strategic possibilities are, according to his interpretation, constituted by one's capacities in terms of technology, finance, and organizational capital.

Two overarching approaches are presented but, as we show, individual contributions also considerably differ from one another. Further confusion apparently comes from the mismatch in their terminologies. How does this knowledge help us understand the transformative potential of AI? Since the consensus is that technology itself is not enough to trigger a revolution, we are instructed to search for other components of change. This is a serious challenge, however, because the existing literature raises more questions than it answers. There is no coherent, integrated guidance on how military revolutions originate, what they entail, and how their driving forces interplay. Chapter 2 will bring to light further shortcomings. It will focus on the existing conceptualizations of revolutionary technologies and reveal their limits in light of AI. It is for this reason that we intentionally keep from discussing it here.

Revolutionary outcomes

The assumption that military revolutions are paradigm shifts in the character or conduct of war runs throughout the literature (Krepinevich 1994; Hundley 1999: 9; Gray 2002: 67–68; Sloan 2002: 3; Horowitz 2010: 22–23; Shimko 2010: 9–11). It is accompanied by another assumption that we know as common sense, that the nature of war as organized violence for political ends, as well as its inherent qualities such as friction, fog, ambiguity, chance, uncertainty, and confusion do not change (Gray 1996: 8; Murray 1997; Murray and Knox 2001: 178–179; Sloan 2002: 30; Shimko 2010: 9; Hoffman 2017/2018: 31). However, there is no agreement on how to detect that a complete revolution has taken place. Cohen (1996: 43–51) operationalized it as change in forms of combat, including the fundamental relationship between offence and defence, space and time, fire and manoeuvre, as well as in the system of military organization, the nature of command, and the relative balance of power. Elsewhere, he (2004: 403) recommended to consider it a revolution if military forces, battle processes, and outcomes look fundamentally different. Rogers (2000: 22) assumed that a revolution manifests itself as 'the ease with which "participating" armed forces can defeat "non-participating" ones'. According to van Creveld (1991: 14–15), it is change in the causes and goals of war, blows with which campaigns open and victories with which they end, modes of organization, command and leadership, methods of planning, preparation, execution and evaluation, strategies and tactics, operations and missions, capabilities, conceptual frameworks employed to think about war, and relationships between armed forces and societies. Generalization is barely possible. Since failed or incomplete military revolutions are rarely studied, we also face the problem of false positives (Hundley 1999: 26).

Efforts to theorize change within the military domain and outside of it add to the existing confusion. Some scholars claim that military revolutions can touch upon only a certain number of and not all branches, domains, and functions of warfare (Hundley 1999: 9; Lambeth 2000: 297; Shimko 2010: 9). Others associate them also with changes in the nature of states and societies, the capacity of states to create and project military power, and the global balance of power (Rogers 2000: 26–29; Murray and Knox 2001: 7). Latham (2002: 235 and 262) went further to distinguish three consecutive layers of the overall structure of organized political violence. These are the dominant warfighting paradigm, the social mode of warfare, and the politico-cultural institution of war. His principled argument was that change needs to be analysed accordingly, from short time-spans to long-run transformative trends.

Apart from the nature of change, there have been attempts to theorize the systemic effects of military innovation and diffusion. There exists a broad consensus that military innovations, including new technologies, do not diffuse across states and their militaries in the same way or at the same time (Krause 1992: 206; Toffler and Toffler 1993: 94–96; Goldman and Andres 1999: 80–81 and 122–124; Hundley

1999: 13 and 33; Rogers 2000: 22 and 27; Boot 2006: 26–28; Gray 2006: 11–13; Adamsky 2010: 4; Horowitz 2010: 18–19, 25, and 41–55). Nonetheless, RMA is often portrayed as a holistic, structural, or universal condition (Fitzsimonds and van Tol 1994: 25; Collins and Futter 2015: 2; Jensen 2018: 303). This brings further confusion in regard to outcomes one ought to expect from a military revolution.

Guided by the lowest common denominator, we must anticipate and prepare for a radical change in the character or conduct of war in the light of AI. But we are not given tools for tracking the course and attesting to the completion of the AI-RMA. There are simply no clear and commonly recognized indicators of a full-scale military revolution. What is more, the existing literature provides rather vague and disjointed guidance on all possible tiers and layers of revolutionary change. We must keep in mind that change can manifest at the global level, the state level, and even the level of individual militaries. We are also instructed to search for elements of change in different dimensions of warfare and broader social and political structures. However, there is no integrated framework that would make it possible from a practical point of view.

Historical dynamics

There is a strong habitual tendency to see the history of military revolutions as a sequence of temporal frames where one replaces another and so on (Latham 2002: 234; Liaropoulos 2006: 366). Given the variety of approaches presented above, different scholars visualized different 'patterns of change' (Gray 2002: 53–55). They identified different kinds and numbers of military revolutions. They also periodized, ordered, and named them in different ways. For example, Krepinevich (1994) proposed a set of ten military revolutions. Four ages of technological and military transformation were identified by van Creveld (1991). Boot (2006) also recognized four ages and four respective military revolutions but their nature and composition differed from those of van Creveld (1991). Murray (1997) distinguished four military revolutions too but did it in a different way from Boot (2006). He also discovered a series of less profound transformations, the so-called RMAs. Knox and Murray (2001) came up with five military revolutions, each accompanied by a series of RMAs. Goldman and Andres (1999) recognized 12 revolutionary military innovations. Toffler and Toffler (1993) focused on three major waves of civilizational and military change.

Nevertheless, it is broadly accepted that revolutions in warmaking must be assessed in terms of their magnitude, not speed (Fitzsimonds and van Tol 1994: 25–27; Krepinevich 1994; Rogers 2000: 31; Murray and Knox 2001: 12; Latham 2002: 232; Gray 2006: 11–12; Shimko 2010: 4). On one hand, this is what makes up the common ground. On the other hand, it is yet another point of disagreement. There are many different interpretations of how revolutionary transformations may eventually cumulate or co-exist, underlie, or overlap one another (e.g., Toffler and Toffler 1993: 34–93; Krepinevich 1994; Hallion 1997: 254; Murray 1997; Goldman and Andres 1999: 101; Hundley 1999: 9; Lambeth 2000: 297; Rogers 2000: 22–24; Boot 2006: 38; Shimko 2010: 9–20). There is another way to look at the overall pattern of changes. Rogers (1993) characterized it as 'punctuated equilibrium evolution'. Going beyond the limits of disjointed revolutionary approaches, other scholars described military change as much more gradual and continuous (Falls 1953: 13; Black 1991; Biddle 1998: 5 and 11–32). It is a challenge to determine who gets it right about the ways in which revolutions are interspersed with periods of incremental change.

The question of how the AI-RMA fits in with the respective military history is one of prioritizing one particular approach over others. The choice is not at all easy. Even if we take one approach for granted, we find ourselves in just about the same position. Most of the existing approaches have already introduced a few closely related terms and it is difficult to say precisely how they accommodate or would accommodate AI. It may be hard to delineate the AI-RMA from what we know, as the third and fourth industrial revolutions, the age of automation and the age of autonomous systems, the robotic revolution and the information technology revolution all used to define military change over the last few decades (Worcester 2015: 2; Hoffman 2017/2018: 20; Shaw 2017: 452; Wallace 2018: 42; Um 2019: 1–5). What we need is a more flexible approach going beyond the logic of rigid timeframes.

(R)evolution: Rethinking and reconstructing RMA

To remedy the discerned flaws and guarantee the lasting relevance of RMA, we turn to organization theory, particularly the institutional perspective to social organization. The purpose is not to remedy all of them but to contribute to a more integrated way of thinking about military revolutions. Generally speaking, and here we part with Hannan and Freeman (1989: 3), 'almost all modern collective action takes place in organizational contexts'. What we also find useful is the 'institutional approach'. It emphasizes the role of institutions in the grasp of human action within organizations as social systems (March and Olsen 1998: 948). The analysis that follows brings organizational and institutionalist perspectives together. The word 'organization' is often used interchangeably with the word 'institution', unless explicitly stated otherwise. Both are construed to mean institutionalized organizational structures herein. Our preferred approach allows us to cut through the thickets of disjointed, contradictory, and often incomplete theorizations of RMA. In particular, it gives us grounds to make certain choices, reconcile some differences, and order piecemeal assumptions. The proposed framework is reinforced by insights from biology, sociology, and philosophy, incorporated to enrich our conceptual vocabulary and sharpen the argument. For heuristic purposes, we begin with a robust conceptualization of military affairs. Only then will we proceed to study the phenomenon of RMA.

Military affairs as an institutionalized military order

The notion of military affairs can be grasped through the lens of what organizational theorists call an organizational field (Campbell 2004: 38). The field is put together by a set of organizations jointly constituting 'a common meaning system' (Scott 1995: 56). Institutional logics tend to permeate all field structures (Scott 2014: 225–228), and the field can as a matter of fact be defined as 'a recognized area of institutional life' (DiMaggio and Powell 1983: 148). For this reason, we can call it an institutionalized order too, as also suggested by Bátora and Hynek (2014: 10). Strictly speaking, such an order can be broken down into three levels of complexity: organizational forms embodied either by a single organization or, more often, several organizations engaging in similar activities and utilizing resources along similar patterns; organizational communities represented by functionally integrated systems of interacting organizational forms; and, at a still higher level, organizational ecosystems composed of interacting organizational communities (Baum 1999: 71; Baum and Rao 2004). Some scholars characterize this composite order as a 'nested hierarchy' (Van de Ven and Grazman 1999: 191). Our preferred term is a *multilayered institutionalized order.*

In this book we focus on organizational collectivities, rather than individual organizations or particular organizational forms, because military revolutions usually entail more profound and broad-scale transformations. For heuristic purposes, we divide them into the micro- (organizational communities), meso- (organizational ecosystems), and macro-levels (multilayered institutionalized orders). We conceptualize military affairs as the latter, in particular as an *institutionalized military order*, and heuristically disaggregate it into its constituent elements. According to mainstream theorists of RMA, such constituents are 'militaries' (Fitzsimonds and van Tol 1994) or 'military organisations' (Krepinevich 1994). Our preferred approach allows us to offer a more sophisticated categorization: *military services* (e.g., Army, Navy, Air Force) are regarded as functionally differentiated organizational communities and *national defence architectures* as ecosystems of integrated military services. The proposed differentiation is depicted in Figure 1.1. It is not random. History tells us that revolutions in warfare can have an immediate or more pronounced impact on a particular military service (e.g., the information technology revolution most often associated with the Gulf War and the US Air Force), a particular nation (e.g., Wehrmacht's Blitzkrieg), or the entire world (e.g., the gunpowder revolution).

The relationship between wholes and their parts within such tiered institutionalized orders is one of *supervenience*. For the most part developed by philosophers of mind, this concept implies that wholes are always dependent on their parts but are never reducible to them (Wendt 1999: 155–156). For example, the US Air Force was fundamentally transformed by the information technology revolution gaining pace in 1990s. The American way of war was revolutionized, but could not be reduced solely to their air warfare capability. The logic is the same at a still higher level. Such a transformation of American air power rapidly and profoundly changed the way war was understood and fought by states and violent non-state actors. Indeed, it entailed a major revolutionary transformation in military affairs. However, this revolution did not amount to a worldwide, ubiquitous paradigm shift and a clean break with the other existing warfare paradigms.

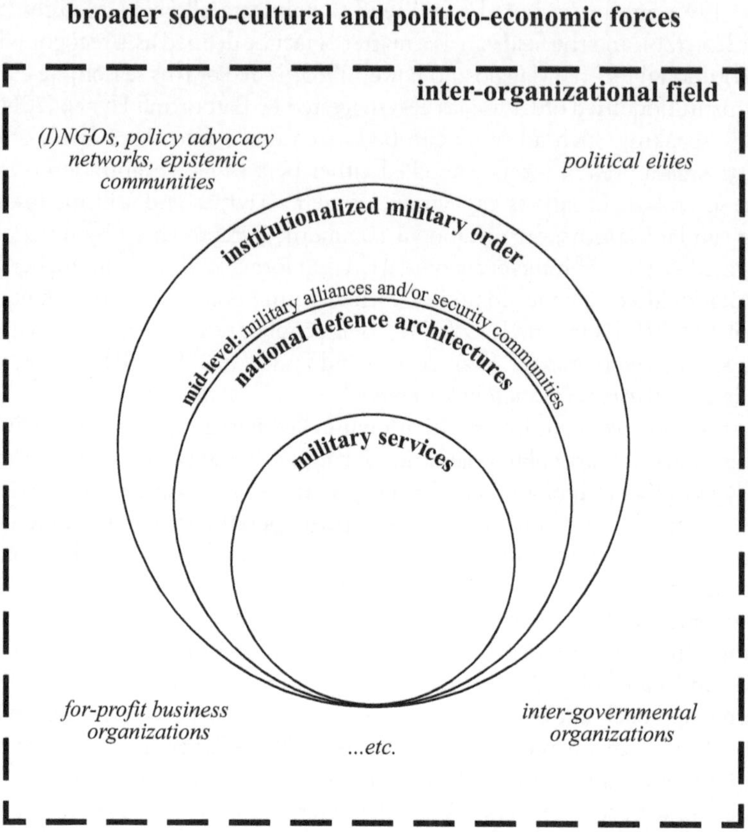

Figure 1.1 Institutionalized military order

There is another point of immense importance. Every institutionalized order is formed in interaction with its environment and continuously engages with it (Bátora and Hynek 2014: 5). The environment encompasses broader – social, cultural, political, economic, etc. – societal forces and other organizational structures (Hannan and Freeman 1989: 7). The latter can be best grasped through the notion of an *inter-organizational field* which implies 'the totality of *relevant* [emphasis added] actors'. The logic of which actors are in and which are out is determined by their involvement in and impact on the respective areas of institutional life (DiMaggio and Powell 1983: 148). For instance, while focusing on healthcare provider organizations (e.g., hospitals), one should take into account the role of purchasers (e.g., individual patients), intermediaries (e.g., commercial insurance companies), and governing bodies (e.g., professional associations) (Scott et al. 2000: 12–13). Figure 1.1 graphically illustrates that the essence of every institutionalized order is therefore inseparable from its immediate context. Chapter 6 will take a broader perspective on military affairs and give an idea of how contextual forces have impacted the course of the AI-RMA. Among them are visions and missions of advocacy groups, epistemic communities, business actors, and political elites. These other sectors of social life have their own structures of organization, which we intentionally refrain from discussing in detail. It is beyond the goals that are set for us in this book. We treat them together here and solely in terms of their immediate or direct relationship to the institutionalized military order.

The aforesaid is still not enough to conceptualize military affairs. To be able to record and track institutional change, we need to identify the so-called 'critical dimensions' that might be subject to change (Campbell 2004: 32). Institutional organization, in principle, consists of a relatively stable set of rules, roles, and regular practices embedded in broader structures of meanings and resources (March and Olsen 1998: 948; Bátora and Hynek 2014: 5). *Military* organization can therefore be defined as a relatively stable set of strategic concepts and doctrines, roles assumed by military and closely related institutions, as well as appropriate operational and training routines. It should simultaneously be seen as embedded in meanings and interpretations provided by political, strategic, geographic, social, cultural, normative, and economic contexts, as well as *available* resources and capabilities, including technologies. While the basic conceptual frame is borrowed from organization theory, the content of this definition is inspired by the existing and aforementioned theorizations of RMA. Our conceptualization is open-ended and allows for the possibility of change in all possible dimensions. It also allows us to observe the variety of institutional organization at different layers and in different contexts. Last but not least, it testifies to the role of technology as an *ever-present contextual parameter*, yet present in a different capacity and to a different degree in all the various forms of military organization (Gray 2006: 23).

Another point, closely connected to the one just mentioned, is that every institutionalized order epitomizes *organizational diversity* (DiMaggio and Powell 1983: 148–149; Hannan and Freeman 1989: 7). Such orders are aggregate but not homogeneous. They encompass multiple, and often conflicting, organizational logics, interpretations, and visions (Bátora and Hynek 2014: 5–7) which may resemble one another only when faced with the same set of contextual conditions (DiMaggio and Powell 1983: 148–149). Latham (2002: 232) specified, referring to Robert Cox, that structures and institutions related to organized political violence can also vary across time and space. It goes without saying that, for example, the American way of war differs sharply from the Afghan way of war or the traditional Vietnamese way. The case of Japan is a good illustration of how one's approach to war can change over time. The country developed a strong 'anti-militarist' orientation after its devastating defeat in World War II. Hashim (1998: 442–443) assumed that it would therefore be reluctant to exploit military revolutions in full. In Chapter 3, we will compare and contrast the US, Russian, and Chinese systems of military organization and respective approaches to the AI-RMA.

Every institutionalized order is also an *interactional order*, and this is what we count as their next distinct feature (Bátora and Hynek 2014: 51). As widely recognized and accurately summarized by Hundley (1999: 13), revolutions in warfare 'frequently bestow an enormous and immediate military advantage on the first nation to exploit them in combat'. However, it is typically a short-lived advantage (Krepinevich 1994; Sloan 2002: 24). This tendency becomes easily comprehensible when seen in an institutional light. Institutional structures co-evolve whilst involved in various forms of interaction such as competition, cooperation, imitation, etc. (Spruyt 1994: 540 and 555; March and Olsen 1998: 955). The history of nuclear weapons and the various strategies and policies that have arisen exemplifies this well. Concerned about the real possibility of the first nuclear bomb being developed in Nazi Germany, the US launched the Manhattan Project in cooperation with the UK and Canada. In theory, revolutionary breakthroughs in one country or a group of countries almost inevitably generate responses, often symmetric or emulative, from other countries (Hashim 1998: 432). Indeed, after the US took a leading role in the coming nuclear revolution, the Soviets immediately aspired to "match" their capabilities (Scott 1992: 182–183). It is a typical example of a symmetric response. We can also think of numerous examples of apparent emulation. In the nineteenth century, many non-European nations tried to learn modern warfare by imitating European armies. Japan was the first non-Western country to successfully emulate the Western way of war (Hashim 1998: 442). The Stanford School's approach to sociological institutionalism made a very interesting and relevant argument in this regard. They challenged the view that formal bureaucratic organizations are 'culture-neutral'. Instead, they stated that the world – in particular *Western* – culture is expanding across and integrating the globe (Finnemore 1996: 328 and 331; Buhari-Gulmez 2010: 256–257). It is true that, in some cases, the diffusion of advanced military technologies is driven primarily by their symbolic qualities and much less by their military utility. Eyre, Suchman, and Alexander (1986: 10) gave an interesting example: 'A stunningly large number of Third World nations [...] maintain only a single "squadron" of four or five fighter aircraft – too few to offer many strategic benefits, but enough to constitute a respectable airshow'. Finnemore (1996: 337) provided another excellent example: 'virtually all states have tripartite military structures, with an army, airforce, and navy – even landlocked states'. These are both signs of cultural affinity with the most technologically and organizationally advanced nations. Saudi arms deals and military procurement is perhaps an even more bizarre example. After its deal with the US in 1979, Saudi Arabia became the second country (after Israel) in the Middle East and in the whole developing world to field an F-15, the best available fighter aircraft. However, the Saudis did not even have the necessary military infrastructure and trained personnel for successfully implementing this modern weaponry (Bents 1995: 95–96). Commenting on their more recent failures, Becca Wasser, a senior policy analyst at the RAND Corporation, reaffirmed: 'Who doesn't love a fancy jet? [...] For the Saudi military, it's been about having prestige items, having a glitter force, without having the skill of being an effective military force' (cited in Cooper 2019). One more form of interaction that is relevant and worth mentioning is mimicry. Borrowed from evolutionary biology, it implies the skill of imitating dominant organizational models to the extent of superficial resemblance (DiMaggio and Powell 1983: 148). One prime example was the development of a nuclear weapons capability under the cover of a civil nuclear energy programme by Iran.

The logic of continuous interaction between and among different institutional structures does not mean that every change will permeate all activities or all countries evenly. This draws our attention to yet another characteristic of most institutionalized orders: they can accommodate both *dominant* and *liminal* patterns, elements, and actors (Bátora and Hynek 2014: 5 and 51–52). What illustrates it well is the exclusive nuclear club. Only a handful of countries have eventually exploited and benefited from the nuclear revolution in military and strategic affairs. Another and less obvious example is the inability of quite a few countries in the Third World to effectively integrate military aviation, a major revolutionary breakthrough of the twentieth century, into their militaries predominantly tailored to land warfare (Cohen 1986: 150–155). There are two potential mechanisms of interaction through which liminalities can be compensated. One is an asymmetric response. It is the deployment of different technologies or the development of new ways of fighting to offset or bypass one's competitive advantage (Hashim 1998: 432). The latter can be illustrated by guerrilla warfare during the Vietnam War. It was a powerful asymmetric response option in light of the US's 'air supremacy' (Mrozek 1998). China's Anti-Access/Area-Denial (A2AD) is an appropriate example of the former. It is a new operational concept and set of technologies designed to hinder the projection of US forces, despite their relative superiority in comparison to Chinese forces, into the Western Pacific. Still one more example, which is unusual and thus worth mentioning, is the Soviet disarmament rhetoric in the early 1950s. It was an effort to buy time and bridge the gap in nuclear weapons and delivery systems with the US (Shoumikhin 2011: 101). The other form of interaction acceptable in such scenarios is external dependency. Extended nuclear deterrence, provided by the US to its European protégés and Japan, is what allowed more countries to enjoy the fruits of the nuclear revolution. NATO's nuclear sharing illustrates the role of alliance politics in the actual transfer of technology and knowledge. Going further than the limiting context of nuclear weapons, we are able to see more clearly that political economy also plays a facilitatory role. Many countries are importers and not producers of revolutionary weapons, for instance advanced conventional weapons (precision-guided munitions, pilotless drones, etc.). Apart from defence cooperation agreements, they can be acquired through commercial arms sales. The provision of not only military equipment but also qualified personnel to operate them or train local forces is yet another form of external dependency. According to recent data, US Forces Korea (USFK) operates about 90 combat planes, 40 attack helicopters, and 60 Patriot missile launchers in South Korea (Reuters 2019). For the past half century, the US has also trained the Saudi military to operate their provided fighter jets and air defence systems (Cooper 2019).

The above provides a practical and thorough conceptualization of military affairs as a *multilayered institutionalized military order*. Its levels of complexity, properties, and qualities are outlined and explained from the perspective of organizational institutionalists. This is a convenient starting point. In the following section, we inquire into the mechanisms and logics of change in such a stratified and highly institutionalized military order. We pose and seek to answer the following questions in order to better understand the phenomenon of RMA. How does change originate? What does it entail? Which layers of institutional organization are involved in or implicated by it? How can we differentiate between revolutionary breakthroughs and more moderate processes of learning and adaptation? How can we understand history in terms of the transformation of the old?

(R)evolution in military affairs: Ontological layers and historical contingencies

This section provides detailed guidance on the content and process of institutional change. Institutional structures exhibit strong inertial tendencies and remain relatively stable over time (Hannan and Freeman 1989: 6; Campbell 2004: 58). However, change is possible and in fact often 'a necessity' (Dekkers 2005: 104). We begin by bringing to light the origins of such change and then proceed to decipher the specificities and outcomes of the very process. The existing literature distinguishes between exogenous and endogenous sources of institutional change (Remuss 2018: 36). The environment itself is one exogenous factor, and change can flow in both directions: institutions adapt to their environment, and vice versa (Hannan and Freeman 1989: 6; March and Olsen 1998: 955). Endogenous forces in turn include all sorts of internal dynamics, including historical trajectories and organizational experimentation (March and Olsen 1998: 955; Bátora and Hynek 2014: 137).

Since institutional change is the result of a mix of exogenous and endogenous factors, it originates from the interplay between *efficiency* and *contingency*. On one hand, it reflects the demands of the changing environment. More efficient organizational forms are created; inefficient ones are dismissed (Hannan and Freeman 1989: 11–13). Spruyt (1994: 554) described it as 'competitive institutional efficiency', echoing the natural selection theory and the principle of the 'survival of the fittest', borrowed from biology (e.g., see Huneman 2015). March and Olsen (1998: 954) chose to use a broader term, namely 'historical efficiency'. On the other hand, organizations can change in response to endogenous and often random processes (Hannan and Freeman 1989: 11–13). Bátora and Hynek (2014: 112) discussed such transformations under the rubrics of 'historical contingency'. These approaches are sometimes considered radically different and even incompatible but we find each, when taken separately, rather naïve. Their joint consideration is the most accurate and reliable way to grasp the complex reality.

The very same logic is pertinent to the issue area of RMA. There are two opposite ways to view military revolutions: the historian's and the strategist's. According to the former, every revolution is a product of exogenous forces such as technological, demographic, or social changes, and it just 'happens'. The latter maintains that revolutions are 'made' and embody solutions, often technological, to a particular country's strategic problems. Shapiro (1999: 136–137) acknowledged that these points of view are radically different but can also be considered as complementary. He argued that exogenous forces 'push' and endogenous forces 'pull' revolutions into being. We generally concur with this argument but also admit that the balance between these two is uneven. In some scenarios, exogenous factors might be decisive; in others, endogenous ones. For example, a revolution in warfare might be a clear consequence of a major social and political upheaval, such as the French *levée en masse* (Fitzsimonds and van Tol 1994: 25). It can, conversely, be a consequence of internal military innovation aimed at solving one's particular strategic problem. What illustrates this well are the US Air-Land Battle (ALB) and NATO's Follow-On Forces Attack (FOFA) which originally pointed towards the information technology revolution (Adamsky 2010: 25–26).

There is an important caveat: contextual forces not only facilitate but can also hinder or at least slow down a successful outcome. For example, nonproliferation commitments have politically and by implication geographically restricted the amplitude of the nuclear revolution. There are some interesting historical examples. In 1139, the Second Lateran Council under Pope Innocent II banned the use of crossbows against Christians (though he granted a dispensation for their use against non-Christians). In the late thirteenth century, the Catholic Church in turn tried to ban all weapons that deployed gunpowder. However, both types of weapons eventually found their way into Christian armies and wars because they were too efficient to be suppressed for ideological reasons (Nolan 2006: 703). The cultural identity of the Ottoman military also hampered the incorporation of firearms into their horsearcher armies in the mid-fifteenth century. The outcome was virtually the same. The Ottomans chose to create an entirely new military organization, the Janissary Corps, to ultimately harness the gunpowder revolution (Steele 2005: 2).

Provided that institutional change is triggered, it brings about non-linear, imperfect, lumpy, tortuous, and disjointed transformations (Campbell 2004: 39; Olsen 2008: 198–200). Institutional change is not a coherent process and there is no single mechanism that underlies it (Olsen 2008: 200). First of all, we assume that it can touch upon any dimensions of institutional organization. Campbell (2004: 39) chose to emphasize that change in one dimension might even cause 'lag' in change in another. This helps explain the role of technology in RMA. We arrive at the conclusion, concurring with Gray (2002: 69), that technologies 'may, but only may, play a vital role'. This perspective accounts for both revolutions driven by technological breakthroughs (e.g., the gunpowder or nuclear revolutions) and revolutions merely buttressed by technologies at hand (e.g., the Blitzkrieg or the Napoleonic Wars). It also allows us to credibly explain why there may be a time 'lag' between the invention of new technologies and their mature application (Cohen 2004: 399). For instance, the very first tank ever deployed on the battlefield was British. However, it alone did not make for a military revolution. It took over 20 years before the German Blitzkrieg produced the desired result. Germany was first to integrate tanks into its armoured formations in a way that, with close air support, revealed their revolutionary potential (Cohen 1996: 46). Evidence suggests that a *tactical* military innovation and a proper, *organizational* military revolution can also be separated by a given time period, or lag (van Nimwegen 2010: 17).

The main question at issue, as spotted by Campbell (2004: 40), is 'how many' dimensions eventually change and 'how fast they do so'. Most of the time, institutions rest in 'equilibrium' and change only 'at the margins'. Such change can be characterized as incremental or, better still, *evolutionary* adaptation. It occurs if 'relatively few' dimensions change at once, or 'if one changes first, then another, and so on, over a long period of time'. Theoretical nuances of learning and adaptation are beyond the scope of this book. Long periods of relative stability are, once in a while, punctuated by abrupt, *revolutionary* transformations. They involve sudden change in 'most, if not all', dimensions of institutional organization. The dichotomy is not as straightforward as it might appear. Institutional change is in reality more complex and often 'a matter of degree'. It can be regarded as more evolutionary in some cases and more revolutionary in others. Perhaps the most eloquent term to describe the general pattern of change is *punctuated evolution* (Campbell 2004: 32–34, 40, and 58). This is a firm analytical basis which allows us to go beyond the limits of different approaches to periodization and track the *intricate* pattern of RMA. Military revolutions are themselves associated with 'radical discontinuity [...] in contrast to linear antecedent behaviour' (Gray 2002: 66). It is no surprise, then, that they are separated by years (e.g., the interwar revolutions in armoured warfare and the Blitzkrieg), decades (e.g., the nuclear revolution of the mid-twentieth century and the information technology revolution of the late twentieth century), or even centuries (e.g., the infantry revolution of the fourteenth century and the artillery revolution of the fifteenth century). The metaphor of 'turning points', borrowed from sociology, most accurately conveys their essence:

> What makes a turning point a turning point rather than a minor ripple is the passage of sufficient time 'on the course' such that it becomes clear that direction has indeed been changed. [...] What makes the trajectories trajectories is their inertial quality, their quality of enduring large amounts of minor variations without any appreciable change in overall direction. [...] An individual actor (biological or social) experiences such a life course as a sequence of trajectories linked to one another via turning points: trajectory, turning point, trajectory, turning point, etc.
>
> (Abbott 2001: 245, 248, and 258)

The best available criteria of radical change are, in our opinion, offered by Kenneth M. Bowler. He operationalized such changes as across-the-board changes in *concepts*, *prescribed courses of action*, and *outcomes* in terms of actual results and costs as well as other intended and unintended social, economic, and political consequences (cited in Albritton 1979: 563). It should also be kept in mind that radical organizational change might involve a recombination of institutional arrangements, their functional reorientation, or a reallocation of resources between and among them (Hannan and Freeman 1989: 8; Remuss 2018: 46). Sometimes new institutional elements and roles are introduced, including through branching (Van de Ven and Grazman 1999: 187 and 200; Remuss 2018: 46). It is also possible that old institutional arrangements are displaced by new ones (Remuss 2018: 46). This takes us back to the importance of institutional *roles*, and any changes therein, too.

This criterion of radical change can be well applied in the military domain. In this sense, a military revolution is a swift, pervasive change *in* strategic concepts and doctrines, the distribution of roles and responsibilities amongst military and closely related institutions, as well as training routines, competitive operational performance, and implications thereof. According to Cohen (2004: 404), among all possible outcomes are also 'battles that never happened, that is, those that were deterred'. The case of nuclear weapons, used in wartime only once, is an excellent example. Some conventional wars have also been deterred by the superiority of more advanced military organizations (e.g., another Arab–Israeli war during the second intifada).

Apart from the very nature of institutional change, we also need to understand the way that it manifests within a multilayered institutionalized order. In view of the aforesaid, we admit that it may be unevenly distributed among the levels of institutional complexity. The general rule is that micro-level processes are more frequent, while macro-level processes produce more dramatic and far-ranging changes (Van de Ven and Grazman 1999: 192). Although some military revolutions have not travelled beyond the borders of a single country, we are more interested in systemic effects of military innovation and diffusion. Wehrmacht's Blitzkrieg is an example of how a minor turning point, yet a turning point nonetheless, can leave larger structures intact. It is also a vivid illustration of how a given revolution might be designed to fit a particular country's strategic needs. However, once in a while, minor turning points line up with other minor turning points and produce a major turning point (Abbott 2001: 257). A very good example is the information technology revolution. Born in the transformation of US airpower, it turned into an impressive revolutionary force penetrating other military services (e.g., the Talon SWORDS is illustrative of its impact on the US army) and many other countries. Organization theory predicts that the magnitude and speed of macro-level changes depend on the responsiveness of the constituent organizations and their ability to absorb change (Hannan and Freeman 1989: 3). Here comes an important remark: not every major change is eventually accommodated by all. We borrow a suitable term from biology to accurately describe this systemic condition: 'frozen evolution'. It implies that few species (here, military organizations) are willing and able to adapt to ever newer demands of the changing environment. Many do not, or simply cannot do so (Flegr 2008: 10, 201, and 205). We can give two examples for illustration: gunpowder, which was itself a revolution, and drones, which are the epitome of the information technology revolution. The former has penetrated virtually every corner of the globe. Although drone technology is still new enough and has spread widely, many countries do not yet have the capacity to develop or effectively integrate it into their defence industries. Another closely related remark is that change can be accommodated to different extents in different contextual situations. What illustrates this well is the adaptation of militaries to motorized and mechanized warfare made possible by the internal combustion engine. Some countries adopted this innovation of the interwar period to marginally improve their military processes (e.g., Italy and Britain). Others developed new military processes and organizations in response (e.g., the Soviet Union and Germany) (Steele 2005: 7–50). In general, change can be observed faster at lower levels of analysis, but takes longer to manifest – and be evaluated accordingly – at higher levels of abstraction (Abbott 2001: 173–174; Campbell 2004: 46). Systemic effects of military innovations depend on each country's understanding of whether its armed forces can exploit them and will benefit from their adoption (Hashim 1998: 431). This assumption makes the most common association of military revolutions with the magnitude, not rapidity, of change meaningful on theoretical grounds. It also helps explain how military revolutions can eventually overlap each other. For example, the nuclear revolution was still underway, with more countries attempting to harness it, when the information technology revolution unfolded in the late twentieth century.

As it derives from the discussion presented above, change is layered in time as well. We reiterate that there are two angles from which layers of history in the realm of military revolutions can be conceived. First, such revolutions may entail transformations in some dimensions of warfare and not others (Cohen 1996: 51; Shimko 2010: 9). Second, they may well be reflected in changes unevenly distributed among and within national defence architectures (Goldman and Andres 1999: 80–81; Lambeth 2000: 297). For that sake, we commit ourselves to Hannan and Freeman's (1989: 20) theoretical position: 'We think that the current diversity of organizational forms reflects the cumulative effect of a long history of variation and selection.'

We come to the conclusion that institutional change, and especially change in military art, features *institutionalized layers driven by a combination of efficiency and tradition, both structured by contingency*. By 'institutionalized layers', we mean two distinct yet inseparable things. From an agency-centred perspective, these are layers of institutional organization depicted in Figure 1.1. From a structural perspective, these are layers of historically contingent arrangements. They manifest themselves as the old is giving way to the new, or the old and the new are being allowed to co-exist. These developments can be best observed at the level of a multilayered institutionalized order. Institutional change can touch upon any dimensions and any layers of institutional organization. It can also begin at any point in time and go on for any amount of time. A wide range of different transformation processes can occur during or near the same period. While it may seem confusing at first glance, our preferred approach brings more flexibility which is perhaps the best tool for a messy reality. It also allows us to treat history as the story of change in its various forms, both evolution and revolution. An important caveat is that contingency does not necessarily mean randomness. It is the central principle of all history because even efficiency is contingent. New elements are coming in while some, but not all, old elements are dying out. Sometimes old elements are applied to new purposes and in new contexts, including as rituals. The constraints of tradition such as path dependency serve as limits to innovation and sometimes even efficiency. They are usually associated with different forms of organizational and cultural inertia, as shown above. Technological change is path dependent in terms of both organizational and technological opportunities, trajectories, and routines. It is the focus of Chapter 2.

Concluding remarks

This chapter dissected and further refined the concept of RMA. The institutional approach to social organization provided an intellectual framework within which disjointed efforts to theorize the phenomenon were integrated. Our motivation was to bridge the gap between the AI-RMA and its *theorization*. Military structures were, first of all, conceptualized as an organizational phenomenon. In doing so, we identified their critical dimensions which *might* be subject to change. This approach gave us a cogent theoretical basis for understanding and analysing organizational, in particular military, change. It also gave us the tools to *contextualize* military organization in a much broader range of organizational, technological, economic, political, social, cultural, and normative processes. Having established a subtle connection between the military and the rest of society, we avoided the dilemma of *whether or not* military change is linked to broader changes in society. Our theoretically informed inference is that these two can be linked but do not have to be. Of particular importance is the role of technology, conceptualized herein as an ever-present contextual parameter which *underlies* military revolutions, at least, and can under certain circumstances *drive* them.

Having distinguished between revolutionary and evolutionary processes to the extent possible, we identified suitable indicators of an *accomplished* military revolution. We also developed a three-level model for grasping the *anatomy* of radical change in the military (Figure 1.1). These and many other aspects, including time lags associated with changes, chains of action and reaction, diverse experiences and contexts, liminal features, and 'frozen' species in the dominant order were all theoretically substantiated within a single intellectual framework.

Our focus in this chapter was on the dynamics of change, with a particular interest in radical change, at different levels of military organization. In the following chapter, we will concentrate on flaws in the existing conceptualizations of revolutionary technologies themselves. We will push further the idea that the key to understanding military revolutions lies, inter alia, in understanding evolution. Chapter 3 will then use the sets of criteria and theoretical assumptions from both chapters to inquire *into* the AI-RMA.

References

Abbott A (2001) *Time Matters: On Theory and Method*. Chicago: University of Chicago Press.

Adamsky D (2010) *The Culture of Military Innovation: The Impact of Cultural Factors on the Revolution in Military Affairs in Russia, the US, and Israel*. Stanford: Stanford University Press.

Albritton RB (1979) 'Measuring Public Policy: Impacts of the Supplemental Security Income Program'. *American Journal of Political Science* 23(3): 559–578.

Bátora J, Hynek N (2014) *Fringe Players and the Diplomatic Order: The 'New' Heteronomy*. New York: Palgrave Macmillan.

Baum JAC (1999) 'Organizational Ecology'. Chapter 3. In: Clegg SR, Hardy C (eds) *Studying Organization: Theory and Method*: 71–108. London: SAGE Publications.

Baum JAC, Rao H (2004) 'Evolutionary Dynamics of Organizational Populations and Communities'. Chapter 8. In: Poole MS, Van de Ven AH (eds) *Handbook of Organizational Change and Innovation*: 212–258. Oxford: Oxford University Press.

Bents ER (1995) *The Sale of US Military Aircraft to Saudi Arabia*. M.A. Thesis presented to the Faculty of the Graduate School of the University of Texas at Austin. Available at: https://apps.dtic.mil/sti/pdfs/ADA294714.pdf.

Biddle S (1998) 'The Past as Prologue: Assessing Theories of Future Warfare'. *Security Studies* 8(1): 1–74.

Black J (1991) *A Military Revolution? Military Change and European Society, 1550–1800*. London: Macmillan.

Boot M (2006) *War Made New: Technology, Warfare, and the Course of History 1500 to Today*. New York: Gotham Books.

Buhari-Gulmez D (2010) 'Stanford School on Sociological Institutionalism: A Global Cultural Approach'. *International Political Sociology* 4(3): 253–270.

Campbell JL (2004) *Institutional Change and Globalization*. Princeton: Princeton University Press.

Cave S, Dihal K (2018) 'Ancient Dreams of Intelligent Machines: 3,000 Years of Robots'. *Nature* 559: 473–475.

Cohen EA (1986) 'Distant Battles: Modern War in the Third World'. *International Security* 10(4): 143–171.

Cohen EA (1996) 'A Revolution in Warfare'. *Foreign Affairs* 75(2): 37–54.

Cohen EA (2004) 'Change and Transformation in Military Affairs'. *Journal of Strategic Studies* 27(3): 395–407.

Collins J, Futter A (2015) Editors' Introduction. In: Collins J, Futter A (eds) *Reassessing the Revolution in Military Affairs: Transformation, Evolution, and Lessons Learnt*. Basingstoke: Palgrave Macmillan.

Cooper H. (2019) 'Attacks Expose Flaws in Saudi Arabia's Expensive Military'. *The New York Times*, 19 September 2019. Available at: https://www.nytimes.com/2019/09/19/us/politics/saudi-military-iran.html.

Dekkers R (2005) *(R)Evolution: Organizations and the Dynamics of the Environment*. New York: Springer.

DiMaggio PJ, Powell WW (1983) 'The Iron Cage Revisited: Institutional Isomorphism and Collective Rationality in Organizational Fields'. *American Sociological Review* 48(2): 147–160.

Eyre DP, Suchman MC, Alexander VD (1986) *Military Procurement as Rational Myth: Notes on the Social Construction of Weapons Proliferation*. Paper presented at the Annual Meeting of the American Sociological Association, New York.

Falls C (1953) *A Hundred Years of War*. London: Gerald Duckworth.

Finnemore M (1996) 'Norms, Culture and World Politics: Insights from Sociology's Institutionalism'. *International Organization* 50(2): 325–347.

Fitzsimonds JR, van Tol JM (1994) 'Revolutions in Military Affairs'. *Joint Force Quarterly* 4: 24–31.

Flegr J (2008) *Frozen Evolution: Or, That's Not the Way It Is, Mr. Darwin – Farewell to Selfish Gene*. Prague: Charles University, Faculty of Science.

Gilli A, Gilli M (2016) 'The Diffusion of Drone Warfare? Industrial, Organizational, and Infrastructural Constraints'. *Security Studies* 25(1): 50–84.

Goldman EO, Andres RB (1999) 'Systemic Effects of Military Innovation and Diffusion'. *Security Studies* 8(4): 79–125.

Gray CS (1996) 'Changing Nature of Warfare?' *Naval War College Review* 49(2): 7–22.

Gray CS (2002) *Strategy for Chaos: Revolutions in Military Affairs and the Evidence of History*. London: Frank Cass.

Gray CS (2006) *Recognizing and Understanding Revolutionary Change in Warfare: The Sovereignty of Context*. Carlisle Barracks: Strategic Studies Institute.

Hallaq B, Somer T, Osula AM, Ngo K, Mitchener-Nissen T (2017) 'Artificial Intelligence within the Military Domain and Cyber Warfare'. In: Scanlon M, Le-Khac NA (eds) *Proceedings of the 16th European Conference on Cyber Warfare and Security*, Dublin, June 2017.

Hallion RP (1997) *Storm Over Iraq: Air Power and the Gulf War*. Washington: Smithsonian Institution.

Hannan MT, Freeman J (1989) *Organizational Ecology*. London: Harvard University Press.

Hashim AS (1998) 'The Revolution in Military Affairs Outside the West'. *Journal of International Affairs* 51(2): 431–445.

Hoffman FG (2017/18) 'Will War's Nature Change in the Seventh Military Revolution?' *Parameters* 47(4): 19–31.

Horowitz M (2010) *The Diffusion of Military Power: Causes and Consequences for International Politics*. Princeton: Princeton University Press.

Hundley RO (1999) *Past Revolutions, Future Transformations: What Can the History of Revolutions in Military Affairs Tell Us about Transforming the U.S. Military?* Santa Monica: RAND.

Huneman P (2015) 'Selection'. Chapter 4. In: Heams T, Huneman P, Lecointre G, Silberstein M (eds) *Handbook of Evolutionary Thinking in the Sciences*: 37–76. London: Springer.

Jensen BM (2018) 'The Role of Ideas in Defense Planning: Revisiting the Revolution in Military Affairs'. *Defence Studies* 18(3): 302–317.

Krause K (1992) *Arms and the State: Patterns of Military Production and Trade*. Cambridge: Cambridge University Press.

Krepinevich AF (1994) 'Cavalry to Computer: The Pattern of Military Revolutions'. *The National Interest* 37: 30–42.

Krepinevich AF (2002) *The Military-Technical Revolution: A Preliminary Assessment*. Washington: Center for Strategic and Budgetary Assessments. Available at: https://csbaonline.org/research/publications/the-military-technical-revolution-a-preliminary-assessment/publication/1.

Lambeth BS (2000) *The Transformation of American Air Power*. Ithaca: Cornell University Press.

Latham A (2002) 'Warfare Transformed: A Braudelian Perspective on the "Revolution in Military Affairs"'. *European Journal of International Relations* 8(2): 231–266.

Liaropoulos AN (2006) 'Revolutions in Warfare: Theoretical Paradigms and Historical Evidence – The Napoleonic and First World War Revolutions in Military Affairs'. *The Journal of Military History* 70(2): 363–384.

March JG, Olsen JP (1998) 'The Institutional Dynamics of International Political Orders'. *International Organization* 52(4): 943–969.

Morgan PM (2000) 'The Impact of the Revolution in Military Affairs'. *Journal of Strategic Studies* 23(1): 132–162.

Mrozek DJ (1998) 'Asymmetric Response to American Air Supremacy in Vietnam'. Part I. In: Matthews LJ (ed) *Challenging the United States Symmetrically and Asymmetrically: Can America Be Defeated?* Pennsylvania: Strategic Studies Institute.

Murray W (1997) 'Thinking about Revolutions in Military Affairs'. *Joint Force Quarterly* 16: 69–76.

Murray W, Knox MG (2001) 'Thinking about Revolutions in Warfare'. Chapter 1. In: Knox MG, Murray W (eds) *The Dynamics of Military Revolution, 1300–2050*: 1–14. Cambridge: Cambridge University Press.

Nilsson NJ (2010) *The Quest for Artificial Intelligence: A History of Ideas and Achievements*. Cambridge: Cambridge University Press.

Nolan CJ (2006) *The Age of Wars of Religion, 1000–1650: An Encyclopedia of Global Warfare and Civilization*, Vol. 2. Westport: Greenwood Press.

Olsen JP (2008) 'Explorations in Institutions and Logics of Appropriateness: An Introductory Essay'. Part III. In: March JG (ed) *Explorations in Organizations*. Stanford: Stanford University Press.

Remuss NL (2018) *Theorising Institutional Change: The Impact of the European Integration Process on the Development of Space Activities in Europe*. Doctoral Thesis accepted by the University of Potsdam. Cham: Springer Theses.

Reuters (2019) 'Factbox: U.S. and South Korea's Security Arrangement'. 13 November 2019. Available at: https://www.reuters.com/article/us-southkorea-usa-military-factbox-idUSKBN1XN09I.

Rogers CJ (1993) 'The Military Revolutions of the Hundred Years' War'. *The Journal of Military History* 57(2): 241–278.

Rogers CJ (2000) '"Military Revolutions" and "Revolutions in Military Affairs": A Historian's Perspective'. Chapter 2. In: Gongora T, von Riekhoff H (eds) *Toward a Revolution in Military Affairs?: Defense and Security at the Dawn of the Twenty-First Century*: 21–36. Westport: Greenwood Press.

Scott HF (1992) 'Soviet Military Doctrine in the Nuclear Age, 1945–1985'. Chapter 6. In: Frank WC, Gillette PS (eds) *Soviet Military Doctrine from Lenin to Gorbachev: 1915–1991*: 175–192. Westport: Greenwood Press.

Scott WR (1995) *Institutions and Organizations*. Thousand Oaks: SAGE Publications.

Scott WR (2014) *Institutions and Organizations: Ideas, Interests, and Identities*. London: SAGE Publications.

Scott WR, Ruef M, Mendel PJ, Caronna CA (2000) *Institutional Change and Healthcare Organizations: From Professional Dominance to Managed Care*. Chicago: University of Chicago Press.

Shapiro J (1999) 'Information and War: Is It a Revolution?' Chapter 5. In: Khalilzad ZM, White JP (eds) *Strategic Appraisal: The Changing Role of Information in Warfare*: 113–153. Washington: RAND.

Shaw IGR (2017) 'Robot Wars: US Empire and Geopolitics in the Robotic Age'. *Security Dialogue* 48(5): 451–470.

Shimko KL (2010) *The Iraq Wars and America's Military Revolution*. New York: Cambridge University Press.

Shoumikhin A (2011) 'Nuclear Weapons in Russian Strategy and Doctrine'. Chapter 3. In: Blank SJ (ed) *Russian Nuclear Weapons: Past, Present, and Future*: 99–159. Carlisle: US Army War College, Strategic Studies Institute.

Sloan E (2002) *The Revolution in Military Affairs: Implications for Canada and NATO*. Montreal: McGill-Queen's University Press.

Spruyt H (1994) 'Institutional Selection in International Relations: State Anarchy as Order'. *International Organization* 48(4): 527–557.

Steele B (2005) *Military Reengineering Between the World Wars*. National Defense Research Institute. Available at: https://www.rand.org/content/dam/rand/pubs/monographs/2005/RAND_MG253.pdf.

Thompson MJ (2011) 'Military Revolutions and Revolutions in Military Affairs: Accurate Descriptions of Change or Intellectual Constructs?' *Strata* 3: 82–108.

Toffler A, Toffler H (1993) *War and Anti-War: Making Sense of Today's Global Chaos*. New York: Warner Books.

Um JS (2019) *Drones as Cyber-Physical Systems: Concepts and Applications for the Fourth Industrial Revolution*. Singapore: Springer.

van Creveld M (1991) *Technology and War: From 2000 B.C. to the Present*. New York: Free Press.

Van de Ven AH, Grazman DN (1999) 'Evolution in a Nested Hierarchy: A Genealogy of Twin Cities Health Care Organizations, 1853–1995'. Chapter 11. In: Baum JAC, McKelvey B (eds) *Variations in Organization Science: In Honor of Donald T. Campbell*: 185–212. London: SAGE Publications.

van Nimwegen O (2010) *The Dutch Army and the Military Revolutions, 1588–1688*. Trans. by May A. Woodbridge: Boydell Press.

Wallace R (2018) *Carl von Clausewitz, the Fog of War, and the AI Revolution: The Real World Is Not a Game of Go*. Cham: Springer.

Wendt A (1999) *Social Theory of International Politics*. Cambridge: Cambridge University Press.

Worcester M (2015) 'Autonomous Warfare – A Revolution in Military Affairs'. *ISPSW Series*, Strategy Series: Focus on Defense and International Security, No. 340. Berlin: Institute for Strategic, Political, Security and Economic Consultancy. Available at: https://www.files.ethz.ch/isn/190160/340_Worcester.pdf.

2 Reconstruction

Artificial Intelligence in Multi-Layered Composite Systems

Introduction

This chapter is the logical extension of the previous chapter in that it seeks to provide a deep and nuanced understanding of technological innovation. It reverses the main question posed in Chapter 1. The focus is on whether and how artificial intelligence (AI) *can itself contribute* to the further elaboration and refinement of the existing theorization of revolutions in military affairs (RMA). What becomes the centre of attention here is the *technological parameter* of RMA. Having evaluated the strength of its current conceptualization in light of AI, we offer a novel perspective. As we show, AI reveals certain trends in technological development. Subsequently, these are taken as the basis for a proposed integrated technological model (Figure 2.1). This model makes existing theorizations of revolutionary technologies much more relevant and better fitting to reality. Not only does it capture the nature of technological change, but it also conceptualizes the *link* between technologies and military revolutions.

Chapter 1 sought to distinguish, at least to the extent possible, between two types of change which occur in relation to/within the military: evolutionary change and revolutionary change. The idea presented there was that revolutions form a natural part of the evolutionary process, which can thus be characterized as *(r)evolution*. Here we take this argument a step further. We draw attention to oft-neglected *evolutionary characteristics* of what is otherwise considered to be a revolution and further call into question the relevance of the term 'RMA'. Therefore, we go beyond the binary thinking underlying a lot of the discussion in Chapter 1.

The technological parameter of RMA: Accommodating AI

Though military revolutions tend to be linked to the exploitation of technological possibilities and breakthroughs, the technological parameter of RMA has gained little attention. The available literature is piecemeal and disjointed in providing a clear and full understanding of its nature. This section offers a careful synthesis of distinct contributions and evaluates whether this framework is generally applicable to AI.

DOI: 10.4324/9781003045489-4

It is broadly accepted that military revolutions often reflect the nexus between civilian or commercial and military technologies (Fitzsimonds and van Tol 1994: 28; Cohen 1996: 41–42; Gray 1996: 14; Horowitz 2010: 31). This nexus can be expressed in two different forms: the common body of knowledge underlying both categories and overlaps in their applications, including what is known as *dual-use* (Buzan 1987: 27). Greater stress is traditionally placed on forward leaps in technology (Sloan 2002: 24). However, some revolutions create new capabilities on the basis of enabling technologies, which can be both familiar and mature and not necessarily innovative. It is mainly due to the possible time 'lag' between the appearance of technologies and their innovative military application (Cohen 2004: 399). Further discord arises from the assumed existence of many different forms of revolutionary technologies and technologies somehow involved in military change. It can sometimes be a *single technological innovation* such as the development of gunpowder or atomic energy that sparks a revolution. Even a *single technological artefact* such as the longbow can facilitate a revolutionary change in warfare (e.g., Murray 1997). Alternatively, a revolution can derive from *confluent streams of technological advancements*. It was the case, for example, with the industrial revolution (machine guns, tanks, submarines, military aircraft, etc.) and the information technology revolution (command, control, communications, computers and intelligence, or C4I for short, systems, precision-guided munitions, stealth technologies, etc.) (Hoffman 2017/2018: 20). The Blitzkrieg, a coordinated effort by tanks, motorized infantry, artillery, and aircraft, is an excellent example of how an *integrated set of technological tools and techniques* can serve as the basis for a military revolution (Hundley 1999: 61–62). It can also be an *innovation in systems integration* that is itself regarded as a revolution in warfare. A good example is the invention of the railway and the telegraph, which are perhaps also the best examples of how civilian technologies alone can produce a military revolution. They constitute the transport revolution and the communications revolution respectively, and the so-called 'age of systems' jointly (van Creveld 1991: 15–16 and 27). What illustrates it equally well is the 'system-of-systems' (later known as network-centric warfare). This operational concept implied the integration of sensors, communications, and precision weapons associated with the information technology revolution (Owens 1996). Sometimes a revolutionary change implies *a set of new technological solutions* integrated into existing military systems. Mechanical equipment such as aircraft, itself rooted in the industrial revolution, has been 'informationized' and 'intellectualized' owing to advancements in computer engineering (Zengfu 1997: 286).

There is no clear understanding, let alone integrated theorization, of the technological parameter of RMA. The limits of the existing literature are most clearly exposed in light of AI. AI is such an omni-present and omni-functional technology that it is beyond every possible distinction. The practical utility of the current scholarship, in our view, boils down to one thing at best: AI is indeed at the nexus of civil and military technologies as a dual-use or, better still, *general-purpose* technology (Horowitz 2018: 39 and 50–52). However, it becomes particularly puzzling as to whether we can think of it in terms of familiar categories. AI is essentially an intelligent computer program which mimics human thinking (Gadiyar, Zhang, and Sankaranarayanan 2019: 168). Its revolutionary impact is sometimes compared to gunpowder and nuclear weapons (FLI 2015). We feel that it is possible to plainly characterize it as a specific, well-defined technological breakthrough. Such computer programs are often installed in suitable pieces of hardware, blurring the divide between the physical and the digital and constituting cyber-physical systems (CPS) (Um 2019: 2–3). It makes us think of another dimension to AI: the physical and easily observable. In fact, AI consists of a wide range of technological innovations mainly in computer vision, pattern recognition, speech recognition, natural language processing, machine learning, artificial neural networks, knowledge representation, and reasoning (Nilsson 2010). It *integrates* revolutionary technologies and cannot be reduced to a single technological innovation. Given its complex and aggregative technological nature, AI is sometimes compared to the past analogue of information and communication technologies in terms of its scope (Ayoub and Payne 2016: 808–810). On the battlefield, however, AI will manifest itself in dispersed but integrated teams of intelligent things and humans (Kott 2018: 63–64; Kott and Stump 2019: 48). Since we cannot yet imagine its actual operation outside of such teams, that in itself – especially if it becomes widespread – will have a revolutionary impact. At the same time, AI has a number of innovative features that can help integrate systems and networks in support of battlefield operations. It opens up unprecedented opportunities in the area of battlefield connectivity as it enables 'swarm intelligence' inspired by social insect behaviour (Dehuri, Cho, and Ghosh 2011: 1–2). Notwithstanding all its innovative properties, AI is readily integrated into machines with familiar physical characteristics such as, for instance, drones (Um 2019: 269). There are cases, therefore, in which its role consists of the *intelligentization* of already existing machinery. We develop this point further in this chapter and in Chapter 3.

AI transcends boundaries in several respects. It embodies all the various forms of technological breakthroughs traditionally associated with military revolutions. What this means is that none of them is sufficient to accurately capture its most revolutionary qualities. AI is not an extreme or unique case in this regard. The information technology revolution, as shown above, is a great example of how uneven and multifaceted technological innovations can underlie a military revolution. However, AI goes a step further. It enables technologies themselves to become and be defined accordingly as *agents* and even *collective actors* (Coeckelbergh 2011; Layton 2018: 41). As we will explain in more detail, they will be capable of intelligent decision-making, both individual and collective. The distinction between subjects and objects will become increasingly blurred. This is a new and fairly unique feature of only one kind of technology: those equipped with AI. Their revolutionary potential consists precisely of this.

Finally, AI challenges the distinction between the old and the new as two competing natures of technologies involved in military change. We cannot unequivocally evaluate its degree of novelty or, contrarily, its maturity at the *moment* of its revolutionary impact on warfare. As already indicated, AI is a locus where a large number of technological streams meet. Nilsson (2010) traced out the long history of the quest for AI and proved that its constituent technologies have displayed different rates of development at different time spans. That being the case, AI research and development should be seen as stretching for decades since roughly the mid-twentieth century. Nothing more definite could be learned from the current state of affairs because there is a difference in the sophistication and range of civil and military applications of AI. AI algorithms have firmly penetrated our daily life. For example, they lie at the basis of Netflix, YouTube, Amazon, and iTunes recommendations, Google search results, and voice-controlled personal assistants such as Apple's Siri, Amazon's Alexa, and Yandex's Alice. AI also underlies, among other things, advanced driver assistance systems and self-driving cars, route-finding programs, natural language translation systems, diagnostic systems in medicine, and facial recognition systems at airports, banks, and other places where identities must be verified (Nilsson 2010: 603–633). But its aptitude for military tasks is 'still more theory than reality' (Horowitz, Kahn, and Mahoney 2020: 530). AI is only now *being* tailored for a range of military functions (Sayler 2020: 10–16; Ventre 2020: 143–146).

Towards an integrated technological model: Layers, systems, and domains

The previous section scrutinized the technological parameter of RMA in light of AI. It also opened space for its systematic reconsideration. In this section, we go beyond the existing, and fragmented, analyses of revolutionary technologies and technologies involved in military change. We take a *longue durée* perspective and offer an integrated technological model which outsteps the heuristic value of categories and the question of novelty and maturity (Figure 2.1). In doing so, we focus on *layers of systems capabilities* which have taken shape in the course of three historical processes: mechanization, automation, and autonomization. It is a novel representation and arrangement of what is traditionally conceived as generations of technological change (e.g., van Creveld 1991; Boot 2006). It takes inspiration from the concept of *composite systems*. Borrowed from computer science, this notion implies the increasing complexity of designed systems, themselves composable and decomposable (Levin 1998: 1–5, 2006: 9 and 65–66). This approach also allows us to obtain a better grasp of technological progress in terms of the reliance of future efforts on past achievements (Dosi and Nelson 2013: 16–17). Numerous examples are used to illustrate the relevance of the proposed model in dealing with the complex technological reality and RMA.

Four layers are identified based on the following operationalization of systems capabilities. By 'systems', we mean pieces of either hardware or software, or structured sets of both. No distinction is made between civilian and military technologies. It is because, as previously discussed, they cannot be strictly differentiated in terms of their impact on warfare. The word 'capabilities' requires a more thorough operationalization. First, we distinguish between the main sources of energy or, in other words, the dominant forces of production and destruction. Second, we trace change in the modes of systems operation and integration. With respect to systems operation, we primarily focus on the principles of human–machine interaction. Systems integration is analysed from the perspectives of data transmission and transportation executed either by humans with the help of technological tools or – at a much later period – technological systems themselves. Finally, we draw attention to the changing character of technologies and their ability to provide effects in five warfighting domains: land, sea, air, outer space, and cyberspace.

The analysis that follows still refers to particular historical epochs and the findings in part reflect the classical view of historical change. The point is not that we cannot avoid this but that we *must* follow this logic for certain reasons. It is not at all possible to identify such layers without the knowledge of their historical evolution. We make a difference by going beyond the limits of periodized timeframes and introducing the idea of composite systems to RMA. Technological change is also characterized by a set of evolutionary trends which cannot be completely abandoned. More importantly, they do have some value for understanding AI. This is because they become particularly apparent in its light. We single out three major evolutionary processes and organize our technological model accordingly (Figure 2.1).

Primitive tools

The most primitive technological tools represent *passive materials* employed by humans for achieving their civilian and military goals. The tools derive their energy from biological energy, in particular human and animal muscles. In principle, this category comprises various hand-powered, including animal-empowered, tools (van Creveld 1991: 15, 23–25, and 95).

In civilian life, the operation of such tools primarily consists of hand labour, sometimes extended by animal labor. These are, for example, primitive agrarian tools for working the soil, stone hand-axes, sharp flint blades, daggers, pikes, chariots, and galleys. Warfare-specialized tools belonging to the same category are ballistae, catapults, swords, halberds, maces, javelins, slings, spears, and lances. They are suited to hand-to-hand, and typically face-to-face, combat. There is one important caveat here. Some of these tools have historically displayed overlaps in their military and non-military applications. For instance, galleys could be used for trade, as well as for war and piracy. Spears could be used for combat, as well as for hunting (van Creveld 1991: 22–46 and 65–79; Toffler and Toffler 1993: 39–40). What unites them is that they all depend on roughly the same sources of energy and often also similar raw materials (van Creveld 1991: 27–28). There are two subcategories that stand out most prominently and deserve comment. Some primitive tools are patently more passive than others. They include defensive equipment such as body armour, shields, and helmets, as well as primitive munitions such as crossbow arrows and stone balls adapted for stone-throwing devices. This caveat, although redundant at first sight, allows us to trace the evolution of munitions and compare primitive tools of defence with today's nearly autonomous defence systems. Some other equipment, still in an essentially primitive state, such as chariots, catapults, crane-like devices, long-distance, and standoff weapons, can in fact extend the limits of biological energy. However, they fall under the same category because they are still powered by the muscles of humans and animals (van Creveld 1991: 27–44 and 77–114; Toffler and Toffler 1993: 39).

There was no such thing as systems integration at a time when technologies were nothing more than passive materials. Information flows and the delivery of people, goods, and services to their destinations were managed by humans and relied on their physical effort. Besides the use of human messengers, the most primitive means of communication relied on acoustic equipment such as horns, trumpets, bugles, gongs, cymbals, and bells, as well as visual tools such as flags, smoke, flashing mirrors and fire. Besides the shoulders of men and the backs of animals, early transportation needs were satisfied by oared water transport and wheeled wagons (van Creveld 1991: 29–59 and 66–67; Toffler and Toffler 1993: 39; NRC 1997: 13–15).

There is usually no debate that some primitive tools are suited for well-defined civilian purposes, while others are suited for certain military tasks. It is the principle of *material determinacy*, according to which materials by their own natures determine the character of their use. As illustrated by van Creveld (1991: 33), 'a sword remained a sword, a lance a lance, and a shield – a shield'.

Primitive technological tools are suited for the exploitation of land and sea and, speaking strictly in military terms, for terrestrial and naval warfare (Andress and Winterfeld 2011: 26). The boundary between these two domains is often well identified. However, some military equipment belonging to the same category also allowed for amphibious operations in very early times. History tells us that, for instance, some types of early warships were not intended for naval battles, but were employed in support of ground forces (van Creveld 1991: 22, 38, 64, and 72).

Mechanization

More advanced technological tools derive their energy from non-biological, non-organic, or inanimate sources. Among them are wind, water, gunpowder, coal, steam, oil, electricity, nuclear energy, radio signals, and electromagnetic radiation. More sophisticated energy carriers such as electrical generators, internal combustion engines, and steam engines produce nearly the same effect (van Creveld 1991: 16, 94–95, and 264; Toffler and Toffler 1993: 41; NRC 1997: 16; Boot 2006: 222, 388–390, and 396). The process of mechanization, which we define broadly for the purposes of this book, enabled technologies to *take over parts of the physical job* originally performed by humans (Gupta and Arora 2007: 2).

The operation of such tools relies on mechanized power which saves the use of biological energy and extends biophysical limitations (van Creveld 1991: 95 and 191; Toffler and Toffler 1993: 44). In civilian life, they are exemplified by windmills, waterwheels, water clocks, typewriters, balloons, dirigibles, sailing ships, steam ships, trains, airplanes, and motor vehicles (van Creveld 1991: 56–68, 129, 170, and 197–198; Boot 2006: 413, 280–281, and 296). Instruments of mechanized warfare include cannons, machine guns, rifles, artillery, submarines, tanks, and aircraft (Boot 2006: 549–556). The metallurgical, engineering, and design knowledge and skills necessary to produce typewriters, airplanes, and tractors are much the same as those necessary to produce machine guns, combat aircraft, and tanks (Buzan and Lawson 2015: 249). Mechanized equipment still remains manually operated as it takes part in physical – and by no means mental – labour. Therefore, it plays only a passive role in wartime. Munitions adapted for mechanized tools, including iron and steel based projectiles, are powered by the explosive effect but, in fact, feature passive military materials. It is in only rare cases such as aerial bombs or nuclear warheads that they consist of explosives themselves (van Creveld 1991: 96–99 and 206; Boot 2006: 43). Mechanization also paved the way for early forms of automation, as illustrated by semiautomatic rifles and pressure-activated mines. Automation in these cases is merely mechanical, however, and exists on a smaller scale (Gupta and Arora 2007: 2).

The trend of mechanization gave birth to early forms of systems integration and, as a result, more effective ways of communication and transportation. Radio waves, electrical and wireless telegraphs, telephone cables and radio phones, as well as steam-driven trains running on railways were employed for farther and faster transfers of information, resources, and the labour force (van Creveld 1991: 167 and 170; NRC 1997: 13–16; Waltz 1998: 3; Boot 2006: 161, 280–281, and 295–296). They all relied almost exclusively on manual human labour but allowed for more complex networks (van Creveld 1991: 253). One more caveat is that they were not designed specifically for military use and provide an excellent example of how civilian technologies can satisfy military needs (van Creveld 1991: 325).

In every way comparable to what we expect from primitive tools, mechanized tools feature hard assets and can be easily defined in material terms. Their physical properties, which are typically observable, determine whether they are suitable for use in a particular civil or military process. Even when embedded in networks, they provide well-known and fairly predictable services to assist humans in meeting their needs (Toffler and Toffler 1993: 67; Terrill and Lidstrom 2019: 84).

Aerial and electronic warfare emerged as new and different kinds of combat in previously unexplored environments as a result of mechanization (Boot 2006: 556). The deployment of military aircraft resulted in more complex inter-domain operations. They consisted in the integration of efforts by air and ground, as well as attacks by aviation and naval forces (Toffler and Toffler 1993: 48–63; Boot 2006: 549–556). Amphibious operations persisted but they appeared to play a relatively lesser role vis-à-vis the increased prominence of air-delivered conflict (Boot 2006: 551–552). The trend to more advanced and increasingly wireless communications created the need for the ability to harness the electromagnetic spectrum. This led to confrontations expressed in the form of electronic attacks, electronic surveillance, and electronic countermeasures, most often mobilized in support of operations in the other warfighting domains (van Creveld 1991: 283; Waltz 1998: 3; Frater and Ryan 2001: 12–14; Boot 2006: 417; Anand, Raja, and Rajan 2011: 901).

Automation

Still more sophisticated technologies depend on the generation of new knowledge, itself increasingly dependent upon computing (Toffler and Toffler 1993: 66–67).

Their primary sources of energy include digital information, computer software, and computerized communications (van Creveld 1991: 252–253, 263, and 280; Arquilla and Ronfeldt 1993: 25; Woodford 2006: 6–10). What distinguishes them is that they are equipped with *electronic brains* (Boot 2006: 429). They can *take over parts of the mental job* and save the use of human thinking (Gupta and Arora 2007: 2).

Their operational capabilities are obtained in the process of automation and robotization (van Creveld 1991: 254; Gupta and Arora 2007: 1–3 and 267). Robots are made capable of *sensing* with the help of increasingly varied sensors, *thinking* in a non-mechanical sense, and *acting* with little or no human input (Lin, Bekey, and Abney 2008: 4). Such technologies, same as computer code themselves, are essentially dual-use (Sparrow 2009: 28). Their application for military purposes has manifested in the creation and deployment of precision-guided munitions, advanced air defence systems, as well as unmanned aerial, ground, and underwater vehicles (Sharkey 2008: 14–15; Johnson and Axinn 2013: 130; Horowitz 2016: 90–91). Two particular observations are worth highlighting. In rare cases and mainly for the purposes of immediate defence, human operators are virtually excluded from decision-making. Counter-rocket (e.g., Iron Dome), anti-missile (e.g., MIM-104 Patriot), and anti-aircraft (e.g., S-400 Triumf) systems of the new generation perfectly illustrate this idea (Lele 2017: 59). Another important observation is that new generation munitions have assumed the role of weapons platforms themselves in very specific cases. Loitering munitions such as the IAI Harpy and IAI Harop are perhaps the best examples of this trend (Horowitz 2016: 91; Sehrawat 2017: 40). There is some clarification needed. None of the aforementioned technologies is truly autonomous at the moment. Computer programmers *automate* only certain processes with the help of pre-defined and hand-coded instructions (McFarland 2015: 1327). With this in mind, we can easily distinguish them from next generation technologies discussed below.

The logic of systems integration on the battlefield has changed profoundly in response to these technological innovations. Digital information – be it text, audio, image, or video – travels easily from one computer to another along interconnected computer networks (Woodford 2006: 6–10). The Internet as a carrier of endless amounts of such information is the edifice of the digital world (Boot 2006: 577–578). Given these opportunities, robotic technologies have become embedded in programmable networks (Terrill and Lidstrom 2019: 84). Platform-centric warfare has been replaced by *network-centric warfare*, which allows commanders, sensors, and shooters to synchronize their actions (Anand, Raja, and Rajan 2011: 898). Unmanned systems have also found their way into military logistics. For instance, they are used to resupply cargo (e.g., CQ-10 SnowGoose) or carry supplies for dismounted personnel (e.g., Ground Unmanned Support Surrogate, or GUSS).

Robots are still machines and often look accordingly. What distinguishes them from their precursors discussed above is that they are computer-controlled. Their capacities are linked less to how they look and more to the parameters and complexity of their software (Sparrow 2009: 28; McFarland 2015: 1326–1327). Despite growing in sophistication, hardware is increasingly 'multipurpose' to keep up with rapid advancements in virtual functions (Terrill and Lidstrom 2019: 84). Robotic capabilities can therefore be best measured in terms of electronic characteristics rather than physical assets (Toffler and Toffler 1993: 67; Gray 1996: 16). What is more, computer programs are not always built into hardware and can also give rise to *disembodied* systems. These are not material, though their effects can be tangible and reside within computers themselves (Kott and Stump 2019: 47–48). The principle of *material indeterminacy* has thus come to replace the principle of predictability.

The process of digitalization, automation, and robotization opened the door for the use of outer space as the fourth natural domain for military purposes (Andress and Winterfeld 2011: 26). It also gave birth to the fifth and entirely man-made domain of war – cyberspace (Siroli 2018: 112). In simple terms, this can be seen as a somewhat qualitative extension of electronic warfare (Frater and Ryan 2001: 219). The rise of cyberspace has consisted in the digitalization of warfare in all its environments and, as a result, cross-domain synergies on the integrated battlefield. The electronic dimension of war has come to prevail over the geographic dimension. Traditional geographic distinctions have become increasingly blurred and porous, yet not completely erased (Gray 1996: 18). The relative importance of both cyberspace and outer space vis-à-vis all other domains of war has been in providing multiple communication paths. However, given the increased role of information and communications, both are on the way to becoming full-fledged warfighting domains themselves (Andress and Winterfeld 2011: 26 and 28–29).

Autonomization

AI that mimics human choice and judgement is the enabler of *intelligent computer technologies*. Learning algorithms allow them to detect patterns in data, make decisions, undertake tasks, and dynamically adjust behaviour with no human input. This learning process is called machine learning (ML). Deep learning (DL) is a more advanced subset of ML. It draws on biologically inspired artificial neural networks (ANNs) to enable computer programs to process and learn from larger amounts of multidimensional and imprecise data as well as maintain more functions (Goodfellow, Bengio, and Courville 2016: 95–96 and 151; Gadiyar, Zhang, and Sankaranarayanan 2019: 167–169). Next generation technologies fundamentally differ from the computer technologies discussed above. *Learning algorithms, not computer programmers, define the outputs of their intelligent functioning* (Layton 2018: 7). This is despite the fact that computer programmers do set the initial parameters of such functioning (McFarland 2015: 1327–1329). Ever more sophisticated algorithms make them more effective and efficient in certain tasks than pre-programmed software and even humans (Layton 2018: 9). However, AI has not replicated human intelligence fully (Springer 2018: 9–10). For this reason, we are particularly interested in 'modular', narrow, or weak AI. It concerns machine intelligence which is very specific to particular tasks and contexts and *not* what is known as superintelligence, superhuman intelligence, general, or strong AI (Ayoub and Payne 2016: 795–796). AI technologies go beyond that, however, and also include sophisticated sensors and other reasoning systems (Gadiyar, Zhang, and Sankaranarayanan 2019: 167).

Equipped with AI, intelligent systems are capable of acting autonomously, without human control. Their distinctive feature is that they can learn, at least in some cases, from interactions with their – both physical and digital – environment. In doing so, they acquire knowledge and continuously update their internal models of the world and behavioural algorithms (Layton 2018: 6). All of this is achieved, inter alia, by extracting and interpreting miscellaneous sensory information and sensor fusion. The latter implies the imitation of what the human brain does with different pieces of information gathered by the senses (Hinton 2007: 428; Sehrawat 2017: 41). AI is the apogee of research in computing and robotics. It brings the relatively simple robots discussed above to a much higher level: it enables them to *sense, reason, act, and adapt* (Gadiyar, Zhang, and Sankaranarayanan 2019: 167). The autonomy which it grants to military systems is sometimes measured in terms of their ability to *observe, orient, decide, act (and assess)* – the so-called OODA(A) loop (Hilger 2015: 80; McFarland 2015: 1324–1325). AI technologies are dual-use or, to be more exact, general-purpose (Horowitz 2018: 39 and 50–52). They are increasingly explored and put to use for different civilian and military purposes. In civilian life, they underlie recommendations on e-commerce websites, route-finding programs, content filtering on social networks, and the development of self-driving cars (Nilsson 2010: 615–621). In the military context, they are being incorporated in combat, command and control, logistics, intelligence gathering, surveillance and reconnaissance missions, cyberspace, and information operations (Sayler 2020: 9–16).

AI paves the way for a new and previously unthinkable kind of systems integration. It is a shift towards self-formed, self-organized, and self-contained networks (Kott 2018: 60; Terrill and Lidstrom 2019: 84). Their success hinges on research and development in the field of swarm intelligence. Swarm intelligence algorithms simulate the collective behaviour often observed in social insects (e.g., colonies of ants, bees, or wasps) (Dehuri, Cho, and Ghosh 2011: 2). Individual network elements are expected to react to external and internal stimuli as one whole (Terrill and Lidstrom 2019: 84). They may become virtually inseparable, which means atomistic ontologies are being incorporated into collective and distributed agencies (Coeckelbergh 2011). Militaries around the world are investing in the development of swarm technology (Hambling 2021). AI also has the potential to revolutionize transportation. Producers are competing to commercialize driverless cars (e.g., General Motors, Ford, Tesla) and develop autonomous delivery drones (e.g., Amazon, Google, DHL) (McKay et al. 2020: 25). Military engineers are closely monitoring and contributing to the development of autonomous driving technology (Pasztor 2021). Driverless transportation is an important priority in the military sphere because it presents an opportunity to remove service members from unnecessarily risky situations (Holley 2018). The future of warfare lies well beyond current network-centric approaches (Layton 2018: iii).

AI also marks the increasing indeterminacy of the material being. AI models built in increasingly complex software expand the range of possible virtual functions and contribute to the rise of *general-purpose hardware* (Gadiyar, Zhang, and Sankaranarayanan 2019: 170). The physical-digital divide has also become ever more blurred in light of recent advancements in *cyber-physical systems*. These are associated with the Fourth Industrial Revolution and, in particular, AI (Um 2019: 1–3). However, and here we part with Schwab, this revolution might be more accurately characterized as 'a fusion of technologies that is blurring the lines between the physical, digital, and biological spheres' (Schwab 2016).

AI facilitates autonomous intelligent operations distributed across multiple domains. Distinctions between air, sea, land, outer space, and cyberspace are *almost* vanishing (Layton 2018: 41). AI-based cyber-physical systems will be able to act independently in specific geographic environments and autonomously fight their cyber adversaries (Kott and Stump 2019: 61). If next generation hardware allows for shape-shifting and artificial general intelligence is achieved, they might even be able to operate in more than one geographical domain, replicating the full range of human performance (Lin, Bekey, and Abney 2008: 20; Layton 2018: 19). AI-empowered cyber systems will be able to fight autonomously, both offensively and defensively, in the virtual domain (Layton 2018: 32). AI-enabled networks will combine autonomous terrestrial, maritime, and aerial systems, space assets, and cyber tools capable of displaying collectively 'intelligent' behaviour (Terrill and Lidstrom 2019: 90).

Multi-layered composite systems

The pattern of technological change discussed above exhibits three evolutionary trends which become particularly apparent in light of AI. First, the degree of autonomy expressed in the actions and interactions of technologies themselves has been on the increase. This trend is characterized by two concurrent processes. One of them is the decrease in the degree of human input in and control over outcomes produced by the use of technologies. Technologies have evolved over centuries from being passive equipment for manual use to active decision makers. The other is the decrease in the degree of material determinacy. Technological change has consisted in the evolution from primitive tools associated with well-known effects to autonomous technologies designed for virtual – and therefore volatile – services in systems and networks. The boundaries between physical and electronic characteristics, natural and technological systems, civilian and military applications, weapons platforms and munitions, systems, and networks have become blurred.

Second, hybridization of efforts in the five warfighting domains has gradually taken place and resulted in increasingly diverse forms of cross-domain operations. The greatest scientific achievement has been the creation of virtual reality and the Internet, which jointly constitute cyberspace. Traditional geographical boundaries have become indistinct in wars without fronts – or rather with fronts everywhere – carried out simultaneously in both physical and virtual contexts (Cohen 2004: 405). However, cyberspace has accompanied and integrated land, sea, air, and space operations but not entirely eliminated technological, tactical, and operational distinctions among them (Gray 1996: 18).

The third and perhaps less obvious trend is the growing sophistication of composite systems. The sources of such sophistication range between the cumulation of technological progress and disruptive, often radical, challenges to existing technologies (for the notion of 'creative destruction', see Schumpeter 1942; for the idea of epistemological and technological ruptures and their significance, see Kuhn 1962; for the concept of 'disruptive technology', see Bower and Christensen 1995). The latter is discussed in detail and conceptualized in the previous chapter. We will return to this point once more below. The former does not derive directly from the analysis presented above and has, perhaps for this very reason, often been overlooked. What we mean by the cumulation of technological progress is that generational technological change has produced technologies with more and more advanced capabilities. However, such advancements are *grafted* onto previous technological achievements and do not supersede them as the idea of successive generations holds. Our argument goes beyond the more general assumption that future possibilities of technological innovation are conditional on its past realizations (Dosi and Nelson 2013: 14). We view technological progress through the prism of increasingly composite systems in which *new functions naturally and simply extend functions already possessed.* As we will explain, the process through which new layers of functions are grafted onto already existing layers is both an evolutionary and a revolutionary phenomenon.

The process of mechanization gave birth to many different machines which were no longer hand-powered (e.g., the French electric submarines Goubet I and Goubet II, the British K-class steam-propelled submarines, the Soviet T-34 tank powered by a diesel engine, the British recoil-operated Maxim gun, the German anti-personnel S-mine). Nonetheless, they were all operated by humans. Even though it might seem that a landmine 'decides' when to explode, it does not decide where it is placed (Asaro 2008: 51). It means that the human element, the cornerstone of the most primitive tools of ancient times, did not disappear then.

The process of automation has made it possible for technologies to reason and make judgements in a non-mechanical sense. However, most of them are *machines* which merely execute human instructions encoded in software (McFarland 2015: 1326). What is more, they are operated, though often remotely, and controlled by humans. The so-called 'unmanned' vehicles are capable of autonomous navigation and target tracking but are also remotely piloted: a human watches a computer screen and makes the final decision on when and what to fire upon (e.g., the MQ-1 Predator and MQ-9 Reaper drones, or the Talon SWORDS unmanned ground vehicle) (Johnson and Axinn 2013: 130). They can therefore be best characterized as 'uninhabited' rather than 'unmanned' (Leveringhaus 2016: 3). Highly advanced defence systems also follow predetermined rules even in their fully autonomous *mode*. In addition to this, they are positioned, activated, and supervised by humans (e.g., the Phalanx Close-In Weapon System, or CIWS for short, and the Aegis Combat System) (Walsh 2015: 1). Such systems are often defined as fully autonomous but it makes more sense to treat them as 'dumb' robots (Kastan 2013: 53–54). In simple terms, automated functions are built into machines that are already familiar, with the human element also preserved.

AI algorithms allow for the automation of *intelligent* decision-making. However, technologies embracing this function are for the most part *computer-controlled machines*, though more complex and more capable in many ways (McFarland 2015: 1326–1327). The layer of intelligent functions is just injected into machine software, allowing it to mimic certain human brain functions (e.g., the incorporation of an AI-assisted copilot into the U-2; the conversion of F-16 fighter jets into AI-empowered QF-16s; the enhancement of the autonomous sensing capabilities of the MQ-9 Reaper with AI) (Everstine 2020; Hambling 2020; Katwala 2020). The very fact that intelligent machines can make intelligent decisions does not completely eliminate the human element either. Although intelligent machines might seem to have the capacity for 'choice', their decisions are not altogether independent of human control. Since computers run whatever software is installed on them, the behaviour of even the most intelligent machines *originates* 'not in the machines themselves, but in the minds of their developers' (McFarland 2015: 1327–1329). AI-enhanced composite systems are presented in a way that is more practical, with more concrete examples, in Chapter 3.

The aforesaid allows us to go beyond the existing, often linear, representations of military history and put together an integrated technological model (Figure 2.1). Four shaded belts constitute the four respective *layers* of systems capabilities which ultimately form one whole. The whole circle graphically nuances the nature of composite systems with elements of AI. The shade of each belt is lighter towards the centre. This illustrates the decrease in both the degree of human control and the level of predictability of systems behaviour. The circle is also divided into four sectors which represent the four domains of the natural environment.[1] It is possible to see, moving from the margins into the centre, how the geographical dimension of war has broadened as a result of advancements in systems capabilities. The dots are used to demarcate those layers which do not exist on their own but do, as we argue, manifest themselves in composite systems. For instance, aircrafts matured in connection with advancements in machine design but they do not function independently of human operators, at least at the time of writing. A spacecraft typically requires onboard electronics and sensors, however it is in fact an automated machine. It is also either crewed or monitored and controlled by a crew at the ground station. We mean to emphasize that even the most advanced aircraft has some degree of primitive logic; even the most sophisticated spacecraft is a product of mechanization which can also be reduced to rather primitive operations. The fifth domain, i.e., cyberspace, is positioned towards the centre for two reasons. It has itself emerged as a result of advancements in computing. It also has its place at the crossroads of land, sea, air, and outer space because it penetrates and integrates them all in many ways. The *autonomization* of technology reinforces this trend. It is for this reason that the dividing lines between the four geographical environments are *increasingly* blurred towards the centre of the circle. Therefore, parallel hatching lines not only delineate cyberspace but also depict integrated efforts across the five warfighting domains.[2]

This model is normatively neutral. In principle, it constitutes the structure of *available* technologies that can serve as the basis for military revolutions.

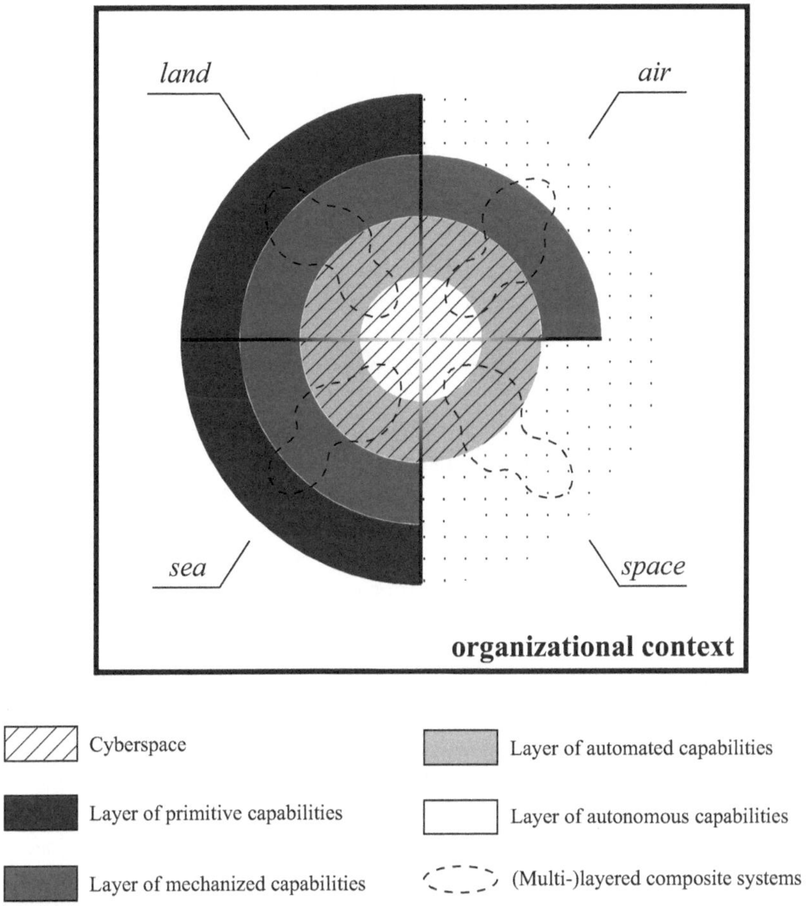

Figure 2.1 Composite systems: Layers of systems capabilities in five warfighting domains

However, it needs to be *contextualized* for being applicable to analyse the respective military history. Chapter 1 conceptualized technologies as part of resources and capabilities *available* to a given military organization, so the composition of what we call the structure of *available* technologies will, in fact, differ from case to case. One's mere possession of technologies also does not mean that their revolutionary potential will be exploited. Many of the existing constraints on the exploitation of technological possibilities and breakthroughs have little to do with available technologies per se but rather with specific organizational arrangements (e.g., institutional inertia, traditional values, civilizational visions, social and legal norms, resource limitations, etc.). Whether a certain transformation can be interpreted so as to constitute a revolution depends on the magnitude of *organizational* change, not necessarily linked to technological change: technological innovations can drive military revolutions, while *available* technologies underlie them at least. (This theory is presented in Chapter 1.) Since military revolutions emerge in the interstices between these two, the structure of *available* technologies should be seen as embedded in the *organizational context* (Figure 2.1).

The model developed in this chapter is primarily designed for grasping the AI-RMA. First of all, it bridges the gap between the ideal typical representations of AI-enhanced composite systems and the messy reality. Therefore, it brings to light *evolutionary characteristics* as well as the *revolutionary potential* of the AI-RMA. The very idea of *layered* composite systems and its visualization captures the former. Their revolutionary potential is determined by a particular set of time-bound, contextual conditions, or *frames*, which the model also incorporates (Figure 2.1). Chapter 3 builds on these assumptions and throws more light on both evolutionary and revolutionary components of the AI-RMA in China, Russia, and the US.

However, the same model can be of help in a wide variety of cases. It captures major technological processes: mechanization, automation, and autonomization. It also consists of a myriad of technological tools with different degrees of sophistication in capabilities. Samples illustrate the logic of AI-enhanced composite systems but it goes without saying that composite systems of lesser complexity are simply hidden from sight (Figure 2.1). Another advantage of the present model is that it accommodates civilian, military, and dual-use technologies. Still more, it is open-ended in a way that it captures both mature and emerging technologies under the rubrics of *available* technologies and *composite* systems. We believe it will prove to be resilient even against future technological innovations. What it all means is that the model bypasses at least three dilemmas: whether technological innovations originate in the military or the civilian world; whether very specific kinds of technologies or combinations of such underlie military revolutions; and whether military revolutions are driven by innovative technologies or innovative applications of established technologies. Technologies themselves can take a variety of forms. Organized around the notion of *systems capabilities*, the model mitigates these differences too and provides a common ground on which a whole lot of different forms can be brought together. Simply put, it can serve as a broader framework to inquire into all kinds of military revolutions which rely on technologies in one way or another. Its mission will be to spotlight the evolutionary characteristics of such technologies and call attention to specific contexts in which they are *made* revolutionary.

Concluding remarks

This chapter leads us to an important conclusion: AI transcends and surpasses all possible forms of revolutionary technologies known to date. With this in mind, we reinterpreted and reconstructed the technological parameter of the RMA. Going beyond a chronological and linear view of history, we identified four layers of capabilities *integrated* into today's advanced *composite systems*. It is how we captured and operationalized the evolutionary characteristics of what is otherwise referred to as the AI-RMA. Having conceptualized the *link* between technological change and opportunities for innovation across military organizations, we also prepared the basis for analysing the revolutionary potential of the AI-RMA. This analysis is reported in Chapter 3. One of our key findings in this chapter, however, is that it is *at least as important* to acknowledge the evolutionary roots of the AI-RMA. Even if AI turns out to have a profound revolutionary impact on the world's militaries, it should not be seen as a *clean* break with the past. This raises a question regarding the extent to which the concept of RMA is an accurate representation of reality. The result of a more general analysis presented here is that it underestimates the evolutionary nature of technologies which, to greater or lesser extents, underlie military revolutions. The following chapter takes a more practical view towards the AI-RMA. It, inter alia, answers the question whether the *rapid* militarization of AI in China, Russia, and the US is indeed a product of evolution, a sign of revolutionary transformation, or both.

Notes

1 Such a division is heuristic. We acknowledge that a particular system can be designed for operation in one domain but have the immediate and most pronounced effects in another domain (e.g. unmanned aerial vehicles used to disable ground-based targets). In this chapter and book, we primarily focus on the former.

2 We recognize that cross-domain applications of technologies have existed long before the emergence of cyberspace. However, it would, in most cases, mean the *adaptation* of specific types of technologies designed for one domain in a different domain. In this chapter, cross-domain operations are understood in relation to two things: omni-use technologies, often borrowed from the civilian world, which *almost readily* fit any of the domains (e.g., computer chips, computer code); and possibilities for the integration of capabilities from *multiple interdependent* domains.

References

Anand D, Raja C, Rajan EG (2011) 'Network Centric Warfare – Concepts and Challenges'. *International Journal of Networking and Communication Engineering* 3(14): 898–902.

Andress J, Winterfeld S (2011) *Cyber Warfare: Techniques, Tactics and Tools for Security Practitioners*. Waltham: Elsevier.

Arquilla J, Ronfeldt D (1993) *Cyberwar Is Coming!* RAND Corporation. Reprint Series. Originally published in Comparative Strategy 12(2).

Asaro P (2008) 'How Just Could a Robot War Be?' Part II. In: Briggle A, Waelbers K, Brey PAE (eds) *Current Issues in Computing and Philosophy*. Amsterdam: IOS Press.

Ayoub K, Payne K (2016) 'Strategy in the Age of Artificial Intelligence'. *Journal of Strategic Studies* 39(5/6): 793–819.

Boot M (2006) *War Made New: Technology, Warfare, and the Course of History 1500 to Today*. New York: Gotham Books.

Bower JL, Christensen CM (1995) 'Disruptive Technology: Catching the Wave'. *Harvard Business Review* 73(1): 43–53.

Buzan B (1987) *An Introduction to Strategic Studies: Military Technology and International Relations*. London: Macmillan Press.

Buzan B, Lawson G (2015) *The Global Transformation: History, Modernity and the Making of International Relations*. Cambridge: Cambridge University Press.

Coeckelbergh M (2011) 'From Killer Machines to Doctrines and Swarms, or Why Ethics of Military Robotics Is Not (Necessarily) about Robots'. *Philosophy and Technology* 24(3): 269–278.

Cohen EA (1996) 'A Revolution in Warfare'. *Foreign Affairs* 75(2): 37–54.

Cohen EA (2004) 'Change and Transformation in Military Affairs'. *Journal of Strategic Studies* 27(3): 395–407.

Dehuri S, Cho SB, Ghosh S (2011) 'Swarm Intelligence and Neural Networks'. Chapter 1. In: Cho SB et al. (eds) *Integration of Swarm Intelligence and Artificial Neural Network*: 1–21. Singapore: World Scientific.

Dosi G, Nelson RR (2013) 'The Evolution of Technologies: An Assessment of the State-of- the-Art'. *Eurasian Business Review* 3(1): 3–46.

Everstine BW (2020) 'U-2 Flies with Artificial Intelligence as Its Co-Pilot'. *Air Force Magazine*, 16 December 2020. Available at: https://www.airforcemag.com/u-2-flies-with-artificial-intelligence-as-its-co-pilot/.

Fitzsimonds JR, van Tol JM (1994) 'Revolutions in Military Affairs'. *Joint Force Quarterly* 4: 24–31.

FLI [Future of Life Institute] (2015) *Autonomous Weapons: An Open Letter from AI and Robotics Researchers*. Announced 28 July 2015. Accessed 3 October 2021. Available at: https://futureoflife.org/open-letter-autonomous-weapons/.

Frater MR, Ryan M (2001) *Electronic Warfare for the Digitized Battlefield*. Norwood: Artech House.

Gadiyar R, Zhang T, Sankaranarayanan A (2019) 'Artificial Intelligence Software and Hardware Platforms'. Chapter 8. In: Gilbert M (ed) *Artificial Intelligence for Autonomous Networks*: 165–188. Boca Raton: CRC Press.

Goodfellow I, Bengio Y, Courville A (2016) *Deep Learning*. Cambridge: MIT Press.

Gray CS (1996) 'The Changing Nature of Warfare?' *Naval War College Review* 49(2): 7–22.

Gupta AK, Arora SK (2007) *Industrial Automation and Robotics*. New Delhi: Laxmi Publications.

Hambling D (2020) 'U.S. to Equip MQ-9 Reaper Drones with Artificial Intelligence'. *Forbes*, 11 December 2020. Available at: https://www.forbes.com/sites/davidhambling/2020/12/11/new-project-will-give-us-mq-9-reaper-drones-artificial-intelligence/?sh=563e78b87a8e.

Hambling D (2021) 'What Are Drone Swarms and Why Does Every Military Suddenly Want One?' *Forbes*, 1 March 2021. Available at: https://www.forbes.com/sites/davidhambling/2021/03/01/what-are-drone-swarms-and-why-does-everyone-suddenly-want-one/?sh=1fbf049c2f5c.

Hilger R (2015) 'A Few Disruptive Thoughts'. Chapter 7. In: Jackson JE (ed) *The U.S. Naval Institute on Naval Innovation*: 79–87. Annapolis: Naval Institute Press.

Hinton GE (2007) 'Learning Multiple Layers of Representation'. *Trends in Cognitive Sciences* 11(10): 428–434.

Hoffman FG (2017/18) 'Will War's Nature Change in the Seventh Military Revolution?' *Parameters* 47(4): 19–31.

Holley P (2018) 'The Military's Latest Plan to Save Lives on the Battlefield: Building Driverless Vehicles'. *The Washington Post*, 2 May 2018. Available at: https://www.washingtonpost.com/news/innovations/wp/2018/05/02/the-militarys-latest-plan-to-save-lives-on-the-battlefield-building-driverless-vehicles/.

Horowitz M (2010) *The Diffusion of Military Power: Causes and Consequences for International Politics*. Princeton: Princeton University Press.

Horowitz M (2018) 'Artificial Intelligence, International Competition, and the Balance of Power'. *Texas National Security Review* 1(3): 37–57.

Horowitz MC (2016) 'Why Words Matter: The Real World Consequences of Defining Autonomous Weapons Systems'. *Temple International and Comparative Law Journal* 30(1): 85–98.

Horowitz MC, Kahn L, Mahoney C (2020) 'The Future of Military Applications of Artificial Intelligence: A Role for Confidence-Building Measures?' *Orbis (Philadelphia)* 64(4): 528–543.

Hundley RO (1999) *Past Revolutions, Future Transformations: What Can the History of Revolutions in Military Affairs Tell Us about Transforming the U.S. Military?* Santa Monica: RAND.

Johnson AM, Axinn S (2013) 'The Morality of Autonomous Robots'. *Journal of Military Ethics* 12(2): 129–141.

Kastan B (2013) 'Autonomous Weapons Systems: A Coming Legal "Singularity"?' *Journal of Law, Technology & Policy* 1: 46–81.

Katwala A (2020) 'The US Air Force Is Turning Old F-16s into Pilotless AI-powered Fighters: Unmanned QF-16s Could be Used to Fly Decoy Routes to Distract from a Manned Aircraft Operating in Stealth Mode'. *Wired*, 27 June 2020. Available at: https://www.wired.co.uk/article/f-16-us-air-force-qf-16.

Kott A (2018) 'Intelligent Autonomous Agents are Key to Cyber Defense of the Future Army Networks'. *The Cyber Defense Review* 3(3): 57–70.

Kott A, Stump E (2019) 'Intelligent Autonomous Things on the Battlefield'. Chapter 3. In: Lawless W et al. (eds) *Artificial Intelligence for the Internet of Everything*: 47–65. London: Academic Press.

Kuhn TS (1962) *The Structure of Scientific Revolutions*. Chicago: University of Chicago Press.

Layton P (2018) *Algorithmic Warfare: Applying Artificial Intelligence to Warfighting*. Canberra: Air Power Development Centre.

Lele A (2017) 'A Military Perspective on Lethal Autonomous Weapon Systems'. In: *Perspectives on Lethal Autonomous Weapon Systems*, UNODA Occasional Papers, No. 30. New York: United Nations Publication.

Leveringhaus A (2016) *Ethics and Autonomous Weapons*. London: Palgrave Macmillan.

Levin MS (1998) *Combinatorial Engineering of Decomposable Systems*. New York: Springer.

Levin MS (2006) *Composite Systems Decisions*. London: Springer.

Lin P, Bekey G, Abney K (2008) *Autonomous Military Robotics: Risk, Ethics, and Design*. Version: 1.0.8. Prepared for US Department of Navy, Office of Naval Research. California Polytechnic State University. Available at: https://digitalcommons.calpoly.edu/cgi/viewcontent.cgi?article=1001&context=phil_fac.

McFarland T (2015) 'Factors Shaping the Legal Implications of Increasingly Autonomous Military Systems'. *International Review of the Red Cross* 97(900): 1313–1339.

McKay S, Boyer ME, Beyene NM, Lerario M, Lewis MW, Stanley KD, Steeb R, Wilson B, Giglio K (2020) *Automating Army Convoys: Technical and Tactical Risks and Opportunities*. Doc. No. RR-2406-A. RAND Corporation. Available at: https://www.rand.org/pubs/research_reports/RR2406.html.

Murray W (1997) 'Thinking about Revolutions in Military Affairs'. *Joint Force Quarterly* 16: 69–76.

Nilsson NJ (2010) *The Quest for Artificial Intelligence: A History of Ideas and Achievements*. Cambridge: Cambridge University Press.

NRC [National Research Council] (1997) *The Evolution of Untethered Communications*. Washington: National Academy Press.

Owens WA (1996) 'The Emerging U.S. System-of-Systems'. *Strategic Forum* 63: 1–6. Doc. No. 20011001042. Institute for National Strategic Studies, National Defense University. Available at: https://apps.dtic.mil/dtic/tr/fulltext/u2/a394313.pdf.

Pasztor A (2021) 'Forget Self-Driving Cars – The Pentagon Wants Autonomous Ships, Choppers and Jets'. *The Wall Street Journal*, 13 February 2021. Available at: https://www.wsj.com/articles/forget-self-driving-carsthe-pentagon-wants-autonomous-ships-choppers-and-jets-11613212200.

Sayler KM (2020) *Artificial Intelligence and National Security*. Originally written by Daniel S. Hoadley. Updated 10 November 2020. Doc. No. R45178. Congressional Research Service. Available at: https://sgp.fas.org/crs/natsec/R45178.pdf.

Schumpeter J (1942) *Capitalism, Socialism and Democracy*. Reprint 1992. London: Routledge.

Schwab K (2016) 'The Fourth Industrial Revolution: What It Means, How to Respond'. *World Economic Forum*, 14 January 2016. Available at: https://www.weforum.org/agenda/2016/01/the-fourth-industrial-revolution-what-it-means-and-how-to-respond/.

Sehrawat V (2017) 'Autonomous Weapon System: Law of Armed Conflict (LOAC) and Other Legal Challenges'. *Computer Law and Security Review* 33(1): 38–56.

Sharkey N (2008) 'Cassandra or False Prophet of Doom: AI Robots and War'. *IEEE Intelligent Systems* 23(4): 14–17.

Siroli GP (2018) 'Considerations on the Cyber Domain as the New Worldwide Battlefield'. *The International Spectator* 53(2): 111–123.

Sloan E (2002) *The Revolution in Military Affairs: Implications for Canada and NATO*. Montreal: McGill-Queen's University Press.

Sparrow R (2009) 'Predators or Plowshares? Arms Control of Robotic Weapons'. *IEEE Technology and Society Magazine* 28(1): 25–29.

Springer PJ (2018) *Outsourcing War to Machines: The Military Robotics Revolution*. Santa Barbara: Praeger.

Terrill S, Lidstrom M (2019) 'Building the Autonomous Networks of the Future'. Chapter 5. In: Gilbert M (ed) *Artificial Intelligence for Autonomous Networks*: 83–100. Boca Raton: CRC Press.

Toffler A, Toffler H (1993) *War and Anti-War: Making Sense of Today's Global Chaos*. New York: Warner Books.

Um JS (2019) *Drones as Cyber-Physical Systems: Concepts and Applications for the Fourth Industrial Revolution*. Singapore: Springer.

van Creveld M (1991) *Technology and War: From 2000 B.C. to the Present*. New York: Free Press.

Ventre D (2020) *Artificial Intelligence, Cybersecurity and Cyber Defense*. London: ISTE Ltd.

Walsh I (2015) 'Political Accountability and Autonomous Weapons'. *Research & Politics* 2(4): 1–6.

Waltz E (1998) *Information Warfare: Principles and Operations*. Norwood: Artech House.

Woodford C (2006) *Digital Technology*. London: Evans Brothers.

Zengfu M (1997) '21st-Century Air Warfare'. Part IV. In: Pillsbury M (ed) *Chinese Views of Future Warfare*. Washington: National Defense University Press.

3 Militarizing Artificial Intelligence in the US, Russia, and China

Introduction

Our intention in this chapter is to analyse existing empirical dynamics and variations of the militarization of artificial intelligence (AI). The goal here is more practical in showing the reader *whether* and *how* the so-called AI revolution in military affairs (AI-RMA) has manifested itself in China, Russia, and the US. Its *revolutionary potential* and *evolutionary characteristics* are examined. In doing so, we bridge the empirical material with concepts and theoretical assumptions developed in the previous two chapters. Case selection is guided by a number of practical concerns. China, Russia, and the US are all great powers with their own competing visions of the world order. All of them manifest global aspirations predicated on geopolitics and geostrategy. The same three are the world's top military spenders and major strategic competitors. Last, but certainly not least, each has openly declared its aspiration for global leadership in the field of AI. The information now available is often crude but still enables us to discern the broad lines of how AI is *becoming* militarized. To provide convincing evidence, the analysis is carried out on two different levels: the level of technologies and the level of military organizations. In regards to the former, we show how *composite systems* – the concept obtained from a *diachronic* analysis of technological change – can be analysed *synchronically*, that is comparatively.

(R)evolutionary technological change

In this section, we analyse the *range* of technologies – designed for different purposes and domains – that are being enhanced by AI. The analysis presented below should be regarded as the continuation of the discussion contained in the preceding chapter. The central conceptual framework employed is that of *composite systems*. Each of these composites introduced below, as we show, contains elements of mechanized equipment, computer hardware and software, as well as AI. Each of them is, at the same time, a *tool* – no matter its degree of sophistication – that serves as an *extension* of the human brain and physical abilities. AI is, therefore, no more than a new form of *human*-machine interaction. It means, inter alia, that virtual technologies which are being utilized *solely* in cyberspace are intentionally

DOI: 10.4324/9781003045489-5

omitted. The focus is on *cyber-physical* composite systems which consist of both physical and software components.

Current information about the most advanced weapons and military technologies in armouries or under development is scarce. For this reason, we take a broader perspective and take into consideration different types of AI technologies with different degrees of sophistication (e.g., machine vision, machine learning, sensor fusion, swarm intelligence, etc.). We take the opportunity to make use of information available from open source reporting, however limited, to draw preliminary conclusions about the nature of the AI-RMA. The language barrier is another challenge but it only concerns the Chinese language. In addition, China has a well-deserved reputation for *opacity* in military developments (Keck 2019). Russian-language sources are covered in addition to English-language ones, which is a considerable advantage for performing an analysis of this type.

The main point of the discussion to follow is that AI is being integrated as a new element into both tested and proven technologies. AI is, in other words, yet another layer of *functions* grafted onto existing, and often mature, layers of systems capabilities and designs. With the help of selected real world examples, we thereby show that evolution and revolution are not mutually exclusive. One caveat is important to note here: we are not aiming for a *cross-country* comparison. Rather, we demonstrate the *range* of composite systems which can be associated with the AI-RMA. Terrestrial, maritime, aerial and space-based systems, military and dual use systems, weapons platforms and munitions, systems independent of each other, and integrated networks are all considered and discussed. Particular attention is paid to their *adaptive* features, whenever applicable, and the *versatile* nature of AI.

United States

In this sub-section, four composite systems developed in the US are examined. The examples are chosen in such a way that each system is designed for a different environment: F-35 (air), Ripsaw M5 (ground), Sea Hunter (surface), Chameleon Constellation (outer space).[1] The selected range of systems also allows us to cover weapons systems (F-35), other military systems (Sea Hunter), and dual-use systems (Chameleon Constellation). It captures both manned (F-35) and unmanned systems (Ripsaw M5).

The **F-35**, a next-generation stealth combat aircraft, is a perfect example of today's composite weapons platforms. There is indeed nothing new about manned aircraft with onboard electronics, sensors, and communication channels. However, new functions will make a sufficiently principled difference between the past generations of piloted aircraft and aircraft such as the F-35. Even though the latter is also designed as a crewed aircraft and has room for one person in the cockpit, the role of a *co-pilot* will be fulfilled by AI (Mizokami 2019a). AI algorithms will take over some of the tasks which would otherwise be assigned to the human pilot. Their role will be, at least, to integrate otherwise disparate pieces of intelligence from a multitude of sensors and other information sources into a single picture for the human pilot. The ability of software algorithms to fuse multi-channel data is known as *sensor fusion* (Osborn 2017; Davis 2021).

This change reflects a more general trend towards the *autonomization* of relatively simple tasks as a way to assist the human pilot. For example, AI has already been tested successfully as a co-pilot on the U-2, one of the oldest aircraft in the US Air Force (Everstine 2020; Roper 2020). In that groundbreaking experiment, AI took over tactical navigation and sensors, while the human pilot could concentrate on flying the plane, authorizing weapons releases, approving changes to flight plans, and communicating with other humans (Mizokami 2020). It may be surprising that one of the oldest aircraft has become one of the first to use AI, but this leads us to an important observation: AI technologies are integratable into different generations of aircraft, as these examples illustrate, and thence *versatile* in their application.

AI will also create new opportunities for *human-machine teaming* (Davis 2021). The F-35 can be accompanied by the so-called loyal wingman, one or more autonomous drones such as the XQ-58A Valkyrie. AI will help the aircraft control such a drone wingman (Osborn 2017; Mizokami 2019a). Loyal wingman-type drones will extend the sensor range and kill radius of the F-35 (Nurkin 2020). Their role will be to scout the route ahead for enemy radars, test and overwhelm enemy air defences, draw fire away from crewed aircraft, scan the skies for aerial threats and pass targeting data on to the crewed aircraft, perform intelligence, surveillance, and reconnaissance missions, and perhaps carry extra weapons (Osborn 2017; Mizokami 2019a, 2021). Valkyrie will itself be equipped with AI. It will be enabled to fly autonomously alongside manned aircraft, learn from prior training missions, stealthily translate between different digital 'languages' and send data between advanced fighter jets such as the F-35 and the F-22, operate in networked autonomous *swarms* and perhaps even leverage support from a much larger swarm of small drones released from its internal weapons bay (Clark 2019; AF.mil 2020; Boulanin et al. 2020: 41; Insinna 2020a, 2020b, 2021; Tirpak 2020; AFRL 2021; Bisht 2021; Newdick 2021; Pawlyk 2021; Wolfe 2021). There is the potential, therefore, that a major air strike could be *led* by a single F-35 (Mizokami 2021). Such technologies are *versatile* too because their development is part of the US Air Force's project 'Skyborg'. This project is set to utilize the different strengths of AI for close interaction and teaming between manned fighter aircraft and unmanned combat aerial vehicles (Mizokami 2019a; Wolfe 2021).

Ripsaw M5, a remote-controlled unmanned battle tank, is another great example of a composite platform developed for the US Army (Randall 2019). Numerous teleoperated robot vehicles, including ground vehicles in particular, have been built and tested in the last few decades. It is only now, however, that their capabilities are being enhanced through retrofitting AI. For example, MQ-9 Reaper, an uncrewed aerial vehicle that has been in service for years, is being equipped with AI. It will be able to carry out autonomous flight, decide where to direct its battery of sensors, and recognize objects on the ground (Hambling 2020). The US Army's modernization priorities are in line with that of the US Air Force. Textron, a leading defence technology company, has, for instance, teamed with software company Shield AI. Their initial collaboration will focus on sharing exploration data between small unmanned aerial systems, developed by Shield AI, and unmanned ground vehicles, developed by Textron. The main goal pursued is to cultivate advanced, *multi-domain* autonomy for a variety of military applications. Shield AI's software will subsequently be integrated into unmanned land, air, and sea systems produced by Textron (IAV 2020; Shield AI 2020). All of this is an excellent illustration of the *cross-domain versatility* of technologies that constitute AI.

Ripsaw M5 is the *fifth generation* of Ripsaw, designed and built by Textron, alongside its subsidiary Howe & Howe and FLIR Systems (IAV 2020). It is the latest in this series of remote-controlled tanks, the first of which was developed from a manned vehicle (Randall 2019). New functions are being interfaced into the existing designs. Ripsaw M5 is capable of 360-degree environment perception and has gained improved situational awareness during both daytime and nighttime (Tadjdeh 2019). AI technologies built into its software eliminate the need for *constant* remote control and allow for some degree of independent operation (Freedberg 2019). Its digital mind is outfitted with machine learning algorithms, essential for building up a comparative picture of the world in real time (Dempsey 2020). Ripsaw M5 will be able to follow another vehicle and operate as a 'wingman' alongside crewed tanks and armoured vehicles, navigate between pre-planned waypoints, serve as a mine roller, plug in a wide variety of sensors and weapons, and serve as a mothership for an airborne drone and a small unmanned ground vehicle which could be dispatched to scout otherwise inaccessible areas (Freedberg 2019; Keller 2019). Ripsaw M5 will be suitable for a variety of settings so it also displays a *versatile design*. It will be adaptable for both surveillance and combat support, and can be fitted with alternative weapons for different challenges (Randall 2019). Its role will be to assist and protect crewed combat vehicles by conducting reconnaissance missions, triggering mines and improvised explosive devices, and providing covering fire (Mizokami 2019b). However, a human will always be in the loop before firing, as highlighted by David Ray, a senior FLIR executive and ex-US Air Force communications expert who served on Air Force One (Dempsey 2020).

Sea Hunter is an anti-submarine surface vessel developed by the Defense Advanced Research Projects Agency (DARPA) and set to complement the US Navy. Its development is part of DARPA's larger initiative to use AI for a wider array of military decisions and tasks (Ghose 2016). Sea Hunter is an intelligence, surveillance, and reconnaissance platform that is not – though has the potential to be – weaponized. This is what distinguishes it from the other two examples presented above. Submarine hunting ships have a long history but this *unmanned* and *autonomous* submarine tracking vessel is endowed with new features for improved capabilities and performance (Naval Technology n.d.). Sea Hunter will not be operated by a full time crew but will run using AI (Pressman 2019). AI, machine learning algorithms, and a variety of sensors, radars, and cameras will allow it to cross thousands of miles of ocean for months without a single crew member aboard, navigate the ocean in accordance with all maritime laws and conventions, collect, analyse, and communicate data in real time, recognize targets, and avoid collisions with other vessels in both day and night conditions (Ghose 2016; Glass 2018; Leidos 2019; Trevithick 2019; Naval Technology n.d.). In doing so, the role of this vessel will be to track enemy submarines in shallow waters, engage in mine countermeasures, and perform intelligence, surveillance, and reconnaissance tasks (Boulanin et al. 2020: 41; Naval Technology n.d.). It will thus take over monotonous and dangerous missions, and lessen the load on manual analytics operators (Leidos 2019).

Sea Hunter is designed to run on its own but humans are still in the loop, at least for now. The vessel can be controlled and guided from a remote station, and even its autonomous operation relies on a *sparse* remote supervisory control model (Ghose 2016; Naval Technology n.d.). Former DARPA Program Manager Scott Littlefield also stressed that these will be humans that will decide on how, when, and where to use such technologies and platforms (Darpa.mil 2016; Deamer 2016). According to open sources, weapons have not been tested on Sea Hunter; however, this vessel has the *capacity* to be armed (Glass 2018). Former Deputy Secretary of Defense Robert Work once raised the possibility of positioning weapons on Sea Hunter (Stewart 2016). *Flexible design* features and earlier indications that such a vessel can be *trained* to abide by international law create a window of opportunity to arm Sea Hunter (Glass 2018). Work stressed, howbeit, same as Former Director of DARPA Steven H. Walker, that lethal decisions will always rest with humans (Stewart 2016; Kreisher 2019).

Chameleon Constellation is another important example of composite technologies that are being outfitted with AI. However, there is a substantial difference between this and the above examples of technologies tailored to military needs. Hypergiant has teamed with the US Air Force to develop a constellation of 24–36 satellites, called Chameleon (Strout 2020). Satellites have long been used for the collection and dissemination of critical imagery and data, both in the civil and military spheres. Hypergiant has taken up this new project to *enhance* autonomous capabilities of satellites with the help of machine learning algorithms and AI. Satellites will themselves be able to determine when repairs are needed and take the necessary action, avoid space debris without instruction, and work together to form an integrated *network* that can communicate with ground stations (Helfrich 2020; Lorek 2020). Project Chameleon draws attention to yet another dimension of *versatile* AI: its *dual purpose*, and this is what makes its applications almost limitless in addition to the aforesaid. One of the project's main goals is to achieve faster and more flexible adaptation of satellite technologies to the present needs (Strout 2020). It is an effort to create a reprogrammable constellation that could be reconfigured through software updates in real-time and retasked for a variety of *both civilian and military needs* while in flight (Helfrich 2020; Lorek 2020). Chameleon is, therefore, a *dual use* configuration with *potential* military applications and itself a perfect illustration of a *versatile design*. The US Air Force and Space Force will be its main users, as revealed by CEO of Hypergiant, Ben Lamm (Lorek 2020).

Russia

This sub-section focuses on four composite systems designed in Russia. The same logic applies to case selection: S-70 Okhotnik (air), Pantsir-SM (ground), and Poseidon (underwater). RB109-A Bylina is chosen as the last system because it operates primarily in the electronic warfare domain so its actual *placement* is less important. What also matters is that the range of systems presented below includes both weapons platforms (e.g., Pantsir-SM) and munitions (e.g., Poseidon).

S-70 Okhotnik, a sixth-generation stealth heavy combat drone, is another suitable example of a composite weapons platform with elements of AI. It is designed to function as a 'loyal wingman' to the Su-57, a fifth-generation stealth fighter developed by Sukhoi for the Russian Air Force (Ria Novosti 2021a). The logic of their close integration is similar to that between the F-35 and the XQ-58A Valkyrie. The latter case was considered above from the perspective of a parent aircraft. Here we discuss in principle the same form of *human-machine teaming* from the perspective of a loyal wingman drone. Recent military history is littered with examples of unmanned aircraft successfully developed and deployed around the world. AI will make a difference, however, and heralds the beginning of a new generation of combat aviation. Okhotnik relies on *autonomous regime* technologies in its ability to take off, complete the task, and return to the airfield (Valagin 2018; TASS 2018a). Even if communications degrade, AI will allow it to determine an appropriate course of action (Melnikov 2019a). The drone will also be able to search for certain types of targets on its own, report on them, and attack (Aksenov 2020). However, it will not be delegated with the *decision* to use weapons (Valagin 2018). It will hit targets *on the command* of the human pilot of the Su-57 (Ria Novosti 2021a). To give the human pilot further assistance, AI will be used to *autonomize* combat surveillance and target acquisition equipment, as well as the flight control system of the Su-57 (Tsargrad 2018).

Pantsir-SM, an advanced air defence complex, is a great example of composite weapons platforms that will boost air combat capabilities of the Russian Armed Forces, and perhaps the Russian Navy. It is developed on the basis of a 'manned' weapons complex and represents one of the latest modifications of Pantsir and, in particular, Pantsir-S1 (Grishchenko 2019; Izvestia 2019; Stepanov 2019; Khrolenko 2020; Ria Novosti 2020). Various air defence complexes have been developed and refined in different parts of the world over the past half century. Pantsir-SM is an illustration of how their functional capabilities can be further enhanced by AI. AI algorithms will allow it to assess the situation and orientate itself, detect and recognize targets, categorize them according to the degree of danger they pose, select the best tactic to repel the attack, and open fire *without human intervention* (Defense World 2020; Khrolenko 2020). Pantsir-SM will strike with absolute precision and, if needed, fire on the move (Ria Novosti 2019; Sharapov 2020). This modernized weapons complex is designed for use against a wide range of airborne targets, including loitering munitions, cruise missiles, ballistic missiles, and drones (Defense World 2020). It will even be able to engage *swarms* of drones, targets flying at hypersonic speed, and multiple targets coming from different directions (Grishchenko 2019; Lenta.Ru 2019; Defense World 2020).

Pantsirs require no crew and no human operators as such (Ramm and Stepovoi 2019). However, they will not be autonomous to the extent that human control is excluded altogether. Pantsir-SM will be mounted on and carried by the multi-purpose truck chassis of the KamAZ 'Tornado' (Stepanov 2019). It will thus not be able to *decide* on its own location, just like other air defence complexes (Walsh 2015: 2). What testifies to its cross-domain applicability and, therefore, *versatile design* is that it can, according to some sources, be installed on ships too (Ria Novosti 2019).

There are still more ways in which AI can reinforce the autonomous capabilities of Pantsir-SM. The Russian Aerospace Forces have, for instance, tested an automated control system with elements of AI. It will integrate different air defence complexes – S-300s, S-400s, Pantsirs – and early warning radars into a single 'armoured fist'. AI-empowered technologies will perform realtime situation analysis and issue recommendations for the use of weapons that are part of this integrated network (Kruglov, Ramm, and Dmitriev 2018).

Poseidon (Status-6), an intercontinental undersea torpedo, elsewhere characterized as an underwater drone and even a mini-submarine, will soon enter into service with the Russian Navy (Pifer 2015; Lockie 2016; Stefanovich 2020: 28). It is perhaps one of the best examples of today's *composite munitions* and deserves careful consideration too. Poseidon illustrates two things at the same time: that mature technologies are being put to new uses, and that innovative technologies are being interfaced into existing designs. It builds upon past advances in nuclear technologies: it can be equipped with a nuclear warhead and, what is more important, it is *powered* by a small nuclear reactor (TASS 2018b; Melnikov 2019b; Edmonds et al. 2021: 92). That means technologies tested and proven on nuclear submarines are being miniaturized and fitted with the design of torpedo tubes. Poseidon will still be the largest and heaviest torpedo in the world (Melnikov 2019b; Baranets 2021).

Poseidon takes these technologies one level up because its complex navigation algorithms and capabilities are *supposed*[2] to be based on AI (Stefanovich 2020: 28; Edmonds et al. 2021: 92 and 124). The torpedo will be able to assess the situation, manoeuvre along the course, and in depth, 'crawl' along the bottom past enemy radars and even lie down on the ground (TASS 2018b; Interfax 2019; Melnikov 2019c; Litovkin 2020; Tulnovskii 2021). Equipped with a nuclear propulsion system, it will have almost unlimited operational time: it will be able to cross oceans and move an intercontinental distance, as well as lie dormant for years close to enemy shores (Melnikov 2019b; Litovkin 2020). Poseidon is also designed as a 'multi-purpose' weapon. It can be used against a broad range of targets, including enemy naval bases, aircraft carriers and ships (both on the high seas and on moorings), industrial facilities, and other infrastructure in the coastal zone (TASS 2018b; Interfax 2019; Baranets 2021). What makes it even more *adaptable* to present needs is that it can be equipped not only with nuclear but also with conventional warheads (TASS 2018b). Some assume that this underwater drone can be used as a situational awareness tool too, as well as for other conventional tasks such as precision minelaying (Stefanovich 2020: 28). Poseidon's independent capabilities are still, at least to some extent, conditional upon human decision. The torpedo will need to be delivered to the launch position by a *crewed* nuclear submarine (Litovkin 2020; Izvestia 2021). Even after it 'sneaks up' on enemy shores, it must wait for an appropriate *command* to hit the target (Baranets 2021).

RB109-A Bylina, one of the latest automated electronic warfare systems, is another interesting example of composite technologies with elements of AI. AI algorithms are used to *fully autonomize* the most complex processes of electronic warfare (Ramm, Litovkin, and Andreev 2017). Bylina will be able to establish communications with electronic warfare stations, higher headquarters, and command posts without human intervention. Among its autonomous functions will also be to analyse the situation in real time, find and recognize different sorts of targets (e.g., enemy radio stations, communication systems, radars, early warning aircraft, satellites), choose the best means to suppress them, give orders to individual electronic warfare stations, and control their operation. What is of particular importance is that it can distinguish with reasonable precision between enemy electronics and one's own (Ramm, Litovkin, and Andreev 2017; Ramm and Stepovoi 2020). Despite its many advanced features, the system in principle *consists* of electronic equipment and communication facilities mounted on high-traffic trucks (Ramm and Stepovoi 2020). Its autonomous capabilities do not mean its complete independence either: human operators will still *control* the process (Ramm, Litovkin, and Andreev 2017).

China

Here we discuss four composite systems produced in China. What guides the selection of cases is their primary operational environment, just as in the previous sub-sections: Blowfish A2 (air), Sharp Claw I (ground), Marine Lizard (surface), and Scavenger Satellites (outer space). Selected examples include those of military (Blowfish A2) and dual-use systems (Scavenger Satellites).

Blowfish A2, a helicopter drone, is yet one more illustration of a composite weapons platform complemented by AI. Helicopter design dates back far into the past. There is also nothing new about remote control technologies and unmanned helicopters. Blowfish A2 is an appropriate example of how these *relatively* mature technologies can be taken one step further. This heli-drone uses computer vision to recognize targets and can perform complex combat missions – such as mid-point or fixed-point detection, fixed-range reconnaissance, and targeted precision strikes – in the autonomous mode (Abadicio 2019; Brar 2019). Such drones can also form a *swarm*, which will itself be able to take off, avoid collision with other aircraft, find the way to a designated target, engage in a coordinated strike, return back to base, and land (Chan 2019; Xuanzun 2019a). AI will guide and coordinate these swarm groups (Peck 2020). At the same time, such drones and swarms are *remote-controlled* and will engage the target once they receive the *order* to attack (Xuanzun 2019a).

Sharp Claw I, a mini unmanned ground combat and reconnaissance platform, is another example which deserves attention. Remote-operated ground vehicles have existed for quite a few years. Sharp Claw I illustrates how their capabilities can be reinforced by the rise of new and innovative technologies that *point towards* AI. The fact that there has been no *explicit* mention of Sharp Claw I being outfitted with AI leads us to another important observation: some systems *can lie midway between* two different layers of systems capabilities (e.g., the layer of automation and tele-operation and that of autonomization and machine intelligence).

Equipped with multiple sensors, machine vision, and night vision, Sharp Claw I is capable of autonomous safe driving, seeking out targets within one kilometre range in all weather, day and night, and returning fire (Army Recognition 2014; Lin and Singer 2014; Strategy Page 2020; Suciu 2020; Thomson 2020). In doing so, it can inspect its surroundings, move in all-terrain conditions and areas of slight slope, cross small ditches, and climb stairs (Army Recognition 2014; Thomson 2020). According to its developers at China North Industries Group Corporation Limited (Norinco), the vehicle is *autonomous* (Suciu 2020). It is designed for use at an active frontline situation and, while stationary, can even act as 'a day and night sentinel' in a combat zone and alert troops of the approaching enemy (Strategy Page 2020). Sharp Claw I is also manoeuverable because of its mini size and can be used to detect and attack targets in buildings, caves, and tunnels (Army Recognition 2014). However, according to some sources, the vehicle is still *remote-controlled*: there is a human operator who supervises the process and makes decisions about the use of weapons (Army Recognition 2014; Lin and Singer 2014; Strategy Page 2020; Thomson 2020).

Marine Lizard (Sea Iguana), the world's first armed amphibious drone boat, provides a further illustration of composite weapons platforms (Defense World 2019; Kania 2020: 4–5). There have been several generations of unmanned surface vehicles, so the principle is not new, but this drone boat is one of the most advanced in the world. Qingdao Wujiang, one of the companies involved in the project, describes itself as an innovative enterprise focused on the development of *deep learning* algorithms, sensors, and control systems for unmanned vehicles. What at least potentially points towards the *versatile* nature of such technologies is that the same enterprise produces other unmanned platforms too (Wood 2019).

Marine Lizard is equipped with advanced navigation, target acquisition, and fire control systems (Wood 2019). AI technologies enable it to sail autonomously, plan routes, avoid obstacles, and swim to shore (Abadicio 2019; Peck 2019; Wood 2019; Xuanzun 2019b). The boat is amphibious because it can be used to seize and defend islands, discover the enemy's positions, and provide covering fire for one's own troops during the assault (Peck 2019; Tang 2019). AI *might*, as some assume, even take over the decision to use weapons (Peck 2019). Such boats can also form a *swarm*, or even an integrated *team* with aerial drones and other robotic warships (*India Times* 2019; Peck 2019). Marine Lizards can, however, be operated by *remote control* at the same time (Abadicio 2019; Defense World 2019; Wood 2019). This can serve as an illustration of *dual control*, i.e., of how the system can be controlled by computer and – potentially – operate in fully autonomous mode and how it can, once required, be switched back to manual mode.

Scavenger Satellites, small satellites with robotic arms, is the last example of composite technologies and, in particular, of mature satellite technologies upgraded through AI. The scavenger program is branded as a space clean-up project which explores the possible uses and benefits of AI (Seidel 2019). Satellites that are being developed as part of this project are all equipped with a triple sensor robotic arm which can recognize the shape of detritus, measure its size and rotation speed, and calculate the exact coupling trajectory with the help of AI (Vecchio 2019). The arm then latches onto the target as it approaches and steers it so that it can burn up while plunging through the atmosphere (Chen 2019). The program is claimed to be peaceful but such satellites may also be utilized for military purposes, especially as most parts of the program are still classified (Chen 2019; Philipp 2019; Vecchio 2019). Being able to grab onto uncooperative targets such as dead spacecraft, rocket casing fragments, and out-of-control space debris, such robotic arms should be able to latch onto still-functioning satellites as well (Chen 2019; Seidel 2019; Vecchio 2019). Their potential uses will, therefore, range from civilian missions such as satellite repair and orbital debris removal to various anti-satellite missions (USCC 2015: 295). The main concern is that the so-called scavenger satellites can instead be used as anti-satellite weapons (ASATs) (Philipp 2019). What offers a decisive military advantage is that such satellites will also be able to remain attached to debris to avoid being tracked from the ground (Chen 2019; Vecchio 2019). All of this is yet another illustration of the *dual use* or, better yet, *versatile* nature of AI.

A revolution in military affairs?

The discussion presented above suggests that AI is the next major technological change which builds on prior technological experience and takes it up one level. AI, as also made clear, brings pervasive changes across virtually all warfare domains. The findings, however, do not necessarily tell us whether the AI-RMA has taken place. To answer that question, we need to understand the degree to which these technological innovations have been institutionalized. Developed in Chapter 1, the indicators of transformation of the entire military organization are as follows: change in strategic concepts and doctrines; change in the distribution of roles and responsibilities among military and closely related institutions; and change in training routines, competitive operational performance, and implications thereof. These changes are examined here using three case studies: China, Russia, and the US. We aim to capture both similarities and context-specific distinct qualities in order to answer both *whether* and *how* the AI-RMA has manifested itself. The research presented in this section is intended to be illustrative, not exhaustive. Selected examples are given to answer the questions posed.

New approaches to technology and warfare

AI has yet to be incorporated into new military doctrines and strategic and operational concepts. This process has begun, but has not gone very far. For this reason, we take a broader perspective on changes in ideas that have taken place in political, military, and academic circles. We also seek to understand these new ideas in a broader context. The transformative impact of AI has been recognized at the level of official discourse in the US. America has already claimed itself to *be* the world leader in AI. The intention is, therefore, to *sustain and enhance* its leadership position (Executive Order 2019). The US has long pioneered the development and application of innovative technologies so this reflects a more general idea of retaining a clear technological edge (Mahnken 2006: 12). In pursuit of this overarching objective, a *whole-of-government approach* to AI innovation has been adopted (Annual Report 2020). The significance of *close collaboration* between the defence and non-defence (academic and commercial) sectors has been recognized (Summary of DoD Strategy 2018: 4 and 7; Gentile et al. 2021: 47–48). This move is yet another step in the creation and maintenance of what President Dwight Eisenhower called the 'military-industrial complex' (Kozyulin 2019: 28).

These and other guiding principles were formulated in the following documents. AI occupied an important place in the US's *Third Offset Strategy* launched in 2014 (Özdemir 2019: 14). One of the core aspects of this document was indeed the realization of the increasing role of civilian technological innovations in meeting military needs (Gentile et al. 2021: 47–48). In 2016, the Obama administration released the *Roadmap for AI* which, inter alia, acknowledged the importance of the global leadership role for the US. Three influential governmental reports were released around the same time: *Preparing for the Future of Artificial Intelligence*, which considered different opportunities, challenges, and possible regulatory responses in view of AI; *Artificial Intelligence, Automation, and the Economy*, which recommended broad strategies that could increase the benefits and mitigate the costs of AI; and *The National Artificial Intelligence Research and Development Strategic Plan*, which established a set of objectives for research and development (R&D) in the field of AI (Özdemir 2019: 14–15). The latter was updated and reissued under the same title in 2019 (National Strategic Plan 2019). The Department of Homeland Security analysed narratives about the benefits and risks associated with AI and published their findings in *Artificial Intelligence Risk to Critical Infrastructure* (2017) (OCIA 2017). AI was for the first time mentioned in the *National Security Strategy* (2017) and the *National Defense Strategy* (2018) (Özdemir 2019: 15). The latter recognized in an explicit manner that new commercial technologies would change the character of war (Summary of NDS 2018: 3). AI once again appeared in the *Interim National Security Strategic Guidance: Renewing America's Advantages* (2021) (INSSG 2021: 8). The Department of Defense (DoD) itself adopted the so-called *AI Strategy: Harnessing AI to Advance Our Security and Prosperity* (2018). The need then arose for service-level annexes to this more general framework and the US Air Force was the first to release its own *AI Strategy* in 2019 (AF.mil 2019; USAF Annex 2019). It is a perfect illustration of the fact that strategic concepts and doctrines can be – and are being – changed at different levels of military organization simultaneously, or near simultaneously. Theoretical aspects related to the multi-layered structure of military organization and the dynamics of change in such structures are contained in Chapter 1. In that same year, 2019,

President Donald Trump launched the so-called *American AI Initiative* which emphasized, among other things, the importance of 'continued' US leadership in AI (OSTP 2019). The *American Artificial Intelligence Initiative: Year One Annual Report* (2020) reviewed progress on implementing the strategic objectives of the Initiative. The US Innovation and Competition Act, motivated by the need to adequately address China's rising technological prowess, was adopted in 2021. It called for greater federal investment in – among other emerging technologies – AI (Zakrzewski 2021).

There is no doubt that a whole new framework is being developed through the exploration of new concepts and principles, but it is clear that it also reflects new understandings of how a new era of conflict will be dominated by AI. AI is, first of all, associated with a broader technological 'revolution' (INSSG 2021: 8). On par with other emerging technologies, it has come to be seen as constitutive of 'the wars of the future' (Summary of NDS 2018: 3). Such wars are often defined in terms of 'algorithmic warfare', i.e., warfare that depends on the performance of algorithms executed by machines (Layton 2018; Crosby 2020; Morgan et al. 2020: 56; Gentile et al. 2021: 47; NSCAI 2021: 77). America's casualty sensitivity is long-standing and its relative risk aversion has been increasingly compensated for by *technology-intensive* approaches to combat (Mahnken 2006: 4; Kriner and Shen 2014: 1174–1175). The American public is, apart from their own military casualties, sensitive to civilian casualties and even enemy military casualties too, which further restrains the freedom of action for political decision makers and military authorities (Sapolsky and Shapiro 1996). The core norms of civil society, acting as a bridge between the individual and the state, are deeply ingrained in America's national identity (Levin 2018). A number of US civil society groups have aligned with the Campaign to Stop Killer Robots and are working to ban fully autonomous weapons (CSKR n.d.). Although the US government does not support the initiative, it has developed a more careful approach towards casualties and machine autonomy. *DoD Directive 3000.09* (2012/2017) requires that all systems be designed to 'allow commanders and operators to exercise appropriate levels of human judgment over the use of force' (Sayler 2020a). AI itself and new forms of combat have thus come to be viewed in terms of human-machine collaboration and 'teaming' (National Strategic Plan 2016: 22, 2019: 14; NSCAI 2021: 34). All of this has taken place against a broader background of mobilizing fundamental American values towards the creation of safe and trustworthy AI (Trumpwhitehouse.archives.gov n.d.).

The Russian state has also recognized the high strategic importance of AI. In 2000, the Russian Ministry of Defence (MoD) adopted an integrated target program called *Robotization of Weapons and Military Equipment – 2015*. In 2015, it proceeded to outline another program, *Creation of Advanced Military Robotics for 2025*, in which the development of unmanned vehicles for different military applications was prioritized. The General Staff of the Russian Armed Forces also developed a plan for the use of robotic systems for military purposes until 2030. In 2016, the Russian government adopted the *Strategy of Scientific and Technological Development of the Russian Federation*. It prioritized, among other things, the development of systems for big data processing, machine learning, and AI (Kozyulin 2019: 25–26). Following the conference *Artificial Intelligence: Problems and Solutions – 2018*, hosted by the Ministry of Defence in collaboration with the Ministry of Science and Higher Education and the Russian Academy of Sciences in March 2018, the Russian government prepared its first formal proposal on AI. It consisted of ten recommendations for advancing AI in Russia. Two months later, President Vladimir Putin issued a decree on the country's national development goals through 2024. The document, inter alia, called for the development of the digital economy, including by means of AI. The government tasked SberBank (now Sber), a bank evolving into a tech company, to produce a roadmap on the development of relevant technologies. Published in 2019, the roadmap focused on technologies such as natural language processing, speech recognition, and computer vision, as well as on how to finance their development. AI-oriented efforts culminated into the *National Strategy on AI Development* (2019). It will serve as the central planning document on AI in Russia through 2030 (Edmonds et al. 2021: 17–20).

At the time of writing, Russia does not have a military-only strategy on AI. However, the Russian state's approach to AI has a strong military dimension and the military sector is one of the strongest when it comes to the development of AI (Slijper, Beck, and Kayser 2019: 16; Markotkin and Chernenko 2020). The significance of AI for the country's 'national security' and 'defence capability' has been formally acknowledged (Kremlin.ru 2019; National Strategy 2019). What is even more important is that the government, and especially the military, *lead* the development of AI in Russia. There is a favourable environment for this approach to be developed. The lack of a 'start-up culture' in Russia leads, unsurprisingly, to the weakness of the private sector (Bendett 2018: 162). The share of private science is, therefore, rather small. The *Strategy of Scientific and Technological Development of the Russian Federation* even admitted that there are no effective programs for the transfer of knowledge and technology between the defence and civilian sectors of the economy in Russia (Presidential Decree 2016: 5; Kozyulin 2019: 28). Russian civil society is, at the same time, not engaged so much on the issue of human control in fully autonomous weapons and has only limited knowledge of the current debate at the Convention on Certain Conventional Weapons (Kozyulin 2019: 26–27).

Such tendencies have their roots. The military has traditionally been perceived as 'an institutional base and legitimizing symbol of Russian statehood and power' (Ermarth 2006: 3–4). Russia, which is surrounded by powerful states and unstable territories and yet lacks physical barriers by which the enemy's access can be hindered, has often found itself in a state of flux. Its 'age-old predicament' has been in defending its vast territory (Graham 2019). The mentality of a 'besieged fortress' has been developed over a long period of time (Igumnova 2011: 264). It has grown into what Kotkin (2016) called 'defensive aggressiveness'. War games, the only true test of one's military superiority other than on the battlefield, have become a tradition. Imperial Russia's techniques and Soviet-era practices testify to their lasting relevance (Caffrey 2019). Deterrence and prevention of aggression are the main goals pursued and, therefore, hold a central place in Moscow's military strategy (Kokoshin, Baluevskii, and Potapov 2015: 17). In March 2020, President Putin explained: 'We are not going to fight against anyone. We are going to create conditions so that nobody wants to fight against us' (TASS 2020).

However, special emphasis is put on the 'general (cross-cutting) character' of AI and the importance of its integration into various sectors of the economy and spheres of social life (National Strategy 2019). Civil-military cooperation and public-private partnerships are encouraged (Presidential Decree 2016: 2; Mil.ru 2018; ЯRobot 2018; National Strategy 2019: 17). Two overarching objectives are pursued in the quest for AI: 'technological sovereignty' (Kremlin.ru 2019; National Strategy 2019: 8) and *messianic* technological leadership. The former is defined as 'self-sufficiency', 'independence', and 'competitiveness' in the realm of science and technology (Presidential Decree 2016: 12; National Strategy 2019). Russia's inclination for sovereignty stems from its traditional adherence to the Westphalian principle of absolute sovereignty (Igumnova 2011: 256). But one caveat is necessary here: the Russian conception of sovereignty relates not only to 'real', i.e., de facto, sovereignty but also to 'reasonable' sovereignty, which is incompatible with autarchy and isolation (Kokoshin 2006). It is perhaps for this reason that *international cooperation* in research and innovation is seen as important, even though it is still discussed in connection with the preservation of the Russian scientific 'identity' (Presidential Decree 2016: 17).

The other objective, as a matter of fact, consists of two different goals: the country's aspiration for leadership, rather 'becoming *one* [emphasis added] of the world leaders' (National Strategy 2019); and a noble argument of Russia willing to share technological know-how (TASS 2017). President Putin has often been cited to claim that the leader in the development of AI would 'rule the world'. It is ignored, however, that he also described it as 'highly undesirable for the monopoly … to be concentrated in one pair of hands' (TASS 2017). Russia's competitive spirit derives from its symbolic demand for great power status (Eitelhuber 2009: 6–8; Igumnova 2011: 256–257 and 262). It is also linked to the fact that Russia perceives itself to be in *permanent* 'struggle' with the West (Adamsky 2018: 52–53; Anno 2019: Chap. 4). The underlying objective is, therefore, to be treated on equal terms in the still mainly Western great power club (Igumnova 2011: 257). Russia has, however, prioritized approaches that substitute for the use of force. The emphasis on *asymmetric*, indirect approaches have deep roots in the Russian military tradition (Adamsky 2018: 49–50). AI-driven asymmetric warfare in the information space, as some believe, might be one of Russia's preferred tools in the age of AI (Polyakova 2018).

Russia's early promise to share technological know-how reflects the country's deep-rooted sense of *messianism* and a special mission. This tendency is traceable to the fall of the Byzantine Empire and the idea of Moscow as the Third Rome. History shows how it has recurred in different situations and contexts: the expansion of religious and imperial tradition (the Imperial period); the spread of communism (the Soviet period); the liberation of the world from the absolute evil of Nazism (the World War II period); and the integration of the former Soviet space under Russian leadership (the post-Cold War period) (Duncan 2000: Chap. 1; Ermarth 2006: 6–7; Igumnova 2011: 257; Domanska 2019: 3).

China's awareness of global trends in military technology and the People's Liberation Army's (PLA's) initial enthusiasm for experimentation underlie China's military initiatives in AI (Kania 2018: 146, 2020: 2). This is a good illustration of both 'push' and 'pull' factors being involved, as theorized in Chapter 1. In China, there is still no coherent consensus on the direction of military AI *within* the PLA, and between the central government, other bureaucratic entities, and the PLA (Ding 2018: 32). Yet certain trends can be identified. China's AI strategy rests on two pillars: the aspiration for *world leadership* and the quest for *self-reliance* in the field of AI. These core principles are contained in *Made in China 2025* (2015) and the *New Generation Artificial Intelligence Development Plan* (2017) (Allen 2019: 3–4). The former is an action plan to further develop China's manufacturing sector, among others, in the field of robotics and AI, by 2025. In the latter document, China declared its intention to become the world leader in AI by 2030 (Özdemir 2019: 19–20). The quest for innovation is an element of the Chinese broader strategy to 'leverage science and technology in pursuit of great power status' (Kania 2020: 2). The following plans and programs have contributed to the same effect: *Robotic Industry Development Plan 2016–2020* (2016) which set concrete goals and strategies in robotics R&D; *'Internet Plus' Artificial Intelligence Three-Year Action Plan 2016–2018* (2016) which called for the creation of infrastructure and platforms for AI; *13th Five-Year Plan for the Development of Strategic Emerging Industries 2016–2020* (2016) which listed AI among the main goals to be pursued by the central government; *13th Five-Year Special Plan for Science and Technology Military-Civil Fusion Special Plan* (2017) which focused on military-civil fusion and the dual use of AI; *Three-Year Action Plan for Promoting Development of a New Generation Artificial Intelligence Industry 2018–2020* (2017), designed to promote the development and in-depth integration of AI (Özdemir 2019: 19–20).

New concepts and organizational principles have come to the fore in political, military, and professional academic circles in China. AI has been referred to as a *disruptive* innovation (Next Generation Plan 2017; Ding 2018: 32). Particular emphasis has been put on 'intelligentized' technologies and the 'intelligentization' of *all-domain* combat capabilities (Kania 2017, 2020: 3; Next Generation Plan 2017; Allen 2019: 5). Writing for *Military Digest*, Ding Ning and Zhang Bing clarified that 'intelligentized' equipment should both possess *intelligence* and be *autonomous* (Fedasiuk 2020: 7). The 'intelligentization' of warfare is in turn associated with a trend towards battlefield 'singularity': since warfare accelerates towards machine speed, the human brain might no longer be able to keep pace with the rhythm of combat and a great part of decision-making power will be given to intelligent machines (Kania 2017, 2018: 143). However, machine intelligence is still defined in terms of *human-machine* 'hybrid' intelligence, the highest form of future intelligence foreseeable at the moment (Kania 2018: 141 and 143).

China applies a 'whole-of-government' approach (Özdemir 2019: 22). Its overarching strategy towards the so-called 'military-civil fusion' will help the PLA to take advantage of AI (Kania 2017). This strategy concerns emerging dual use technologies and aims to both break down barriers and create synergies between economic development and military modernization (Özdemir 2019: 18; Kania and Laskai 2021: 4). China's new national AI strategy has, inter alia, called for the formation of an *all-element, multi-field, two-way* military-civilian integration and the deep integration of AI with the economy, society, and national defence (Next Generation Plan 2017). China has pursued military-civil integration in some form since at least the 1980s. These policies are being *deepened* under the moniker of 'fusion' in the last few years (Kania and Laskai 2021: 6). Marine Lizard is a good example of military-civilian fusion in action: it is under joint development by Wuchuan Shipbuilding Group and Qingdao Wujiang Technology (Wood 2019).

China's main focus is on *command decision making, military deductions* (e.g., war games), and *defence equipment* when it comes to military applications of AI (Next Generation Plan 2017). These priorities reflect the country's distinctive strategic culture and historical experience. The PLA's inclination to consolidate and centralize decision-making and greater vigilance to ensure loyalty to the Chinese Communist Party could lead to high reliance on AI (Kania 2017, 2018: 145–146). Machine intelligence might perhaps be seen as both more reliable and controllable (Kania 2018: 146). AI technologies can also enhance the level of intelligent information processing in the PLA's so-called 'system-of-systems' operations (Burke et al. 2020: 1; Morgan et al. 2020: 61). However, the significance of human-machine interaction and, in particular, the *human element* of command is recognized in the PLA (Kania 2018: 143). China lacks combat experience at the same time and is, therefore, more focused on war games and simulations (Özdemir 2019: 21). The PLA often relies on the lessons learnt from foreign conflicts and tends to evaluate warfare through the lens of military science (Kania 2017, 2018: 146). China's philosophy of *no-first-use* and longer-term tradition of *developing weapons with only the threat of deploying them* is also reflected in its approach to military AI. China has advocated for the ban on the usage, *not* the development and production, of fully autonomous weapons (Roberts et al. 2021: 63–64). Further details on this subject are provided in Chapter 6. More limited, defensive goals will, however, inform the most immediate applications of military AI (Ding 2018: 32). In the last few decades, the principle of *active defence* in local, small-scale wars has gained considerable importance for the PLA (Godwin 1992: 191; Burke et al. 2020: 3). China has, in addition to that, adopted *asymmetric strategies* to challenge the power projection capabilities of the US in the Western Pacific (Ding 2018: 32; Kania 2020: 2; Roberts et al. 2021: 62). That is despite the recent expansion of Beijing's nuclear capabilities, which signalled that power relations between the two competitors are becoming more symmetric (Warrick 2021). The increasing sophistication of unmanned weapons, as discussed in more detail above, might contribute to achieving both goals. Blowfish A2 will, for example, help advance China's *anti-access and area denial capabilities* (A2/AD), i.e., the capacity of its armed forces to block adversaries from traversing its own or disputed territories (Brar 2019). Marine Lizards, which are being developed for the marine defence market, are primarily designed for operations on disputed islands in the South China Sea (Peck 2019; Vavasseur 2019; Wood 2019).

New organizational ecosystems

Transformations at the level of ideas have been accompanied by the emergence of the so-called AI 'ecosystems' in China, Russia, and the US. AI 'ecosystems' refer to – and unite – a wide range of public and private sector partners involved in the innovation process. In theoretical terms, each of them can be best grasped from the perspective of what we called an *inter-organizational field* in Chapter 1. Due to the inherently dual-use nature of AI, with the cutting edge often shifting towards the civilian side, it is important to analyse the organizational complexity of such 'ecosystems'. Below, we single out in a selective manner new institutional arrangements, new organizational principles, new functions and roles of already existing institutions, and other institutional changes that have been introduced. In doing so, we capture the complexity of organizational social contexts and the subsequent diversity in organizations too.

China's AI 'ecosystem' has taken shape in the last few years. The Chinese government has increased their spending on AI by 350 per cent between 2005 and 2015 and is now generally believed to be very close to spending levels currently set in the US. Chinese companies have a great share of the world's total AI start-up funding as well (Özdemir 2019: 18). The PLA's capabilities are being enhanced through the coordinated efforts of military research institutes, commercial enterprises supporting military-civil fusion, and the Chinese defence industry (Kania 2020: 4). The PLA is, at the same time, 'unique' in that it is subordinate to the Chinese Communist Party (Kania 2018: 146). China's so-called 'whole-of-government' approach implies a top-down process in which the government plays a crucial role. The Chinese Communist Party Central Committee controls all the activities, including technological developments. Two institutions are at the forefront of policy making and coordination efforts in science and technological matters: the National Science, Technology and Education Leading Group; and the Chinese Communist Party Central Military Commission. Their duties and responsibilities reflect China's *military-civil fusion* strategy: the former is responsible for civilian applications of technology, while the latter is responsible for military applications. The main private companies, which cooperate with the Chinese government and are responsible for the development of AI, are Baidu, Alibaba, and Tencent (Özdemir 2019: 20 and 22). These are the so-called 'AI national champions' (Roberts et al. 2021: 61). In 2018, China also announced its intention to build a $2 billion AI research park in Beijing. It will be a national-level AI *lab* which will house up to 400 enterprises and serve to attract foreign firms as well (Reuters 2018). All of this illustrates that the Chinese government and the private sector work – and will continue to work – closely together on their shared goals in the field of AI. Such an organizational arrangement makes the application of AI to the military much easier (Özdemir 2019: 19). The Military-Civil Integration Development Commission, established in 2017, is in charge of China's national agenda for *military-civil fusion* (Kania 2017).

The PLA is itself creating mechanisms to *actualize* military-civil integration. One of them is the Military-Civil Integration Intelligent Equipment Research

Institute. Created with support from the PLA, this institute focuses in particular on intelligent robotics, unmanned systems, and AI (Kania 2017). The PLA's continued efforts to implement AI in command and control systems incentivized action by a number of other relevant players. In partnership with Baidu, the 28th Research Institute of the China Electronics Technology Group, a state-owned defence conglomerate, established the Joint Laboratory for Intelligent Command and Control Technologies. Its mission will be to increase the level of intelligentization in command information systems through the introduction of big data and AI. The China Institute of Command and Control, which brings together some of the top experts from the Chinese academia, military, and defence industry, also established a new Intelligent Command and Control Systems Engineering Specialist Committee (Kania 2018: 142–143 and 145).

Russia's overall funding for AI R&D remains comparatively low, even though it is quite generous by their standards (Edmonds et al. 2021: 35). Russia's state-led AI 'ecosystem' consists of the government and the military, which have both taken the lead, state-owned companies, and the growing private sector, as well as the nation's universities (Bendett 2018: 162). State research institutes employ a significant majority of all Russian researchers and account for the largest share of domestic R&D expenses (Kozyulin 2019: 28). Three universities – Moscow State University, Moscow Institute of Physics and Technology, and Moscow Higher School of Economics – are at the forefront of AI-related research activities (Tadjdeh 2021). Russia's most innovative projects in the military sphere fall under the auspices of the Russian MoD and its affiliate institutions. The role of the Advanced Research Foundation, analogous to the US's DARPA, is important in particular (Bendett 2018: 163). Some other organizations are also responsible for military aspects related to the development and application of AI: the MoD Commission for the Development of Robotic Systems for Military Purposes, headed by Russian Defence Minister Sergei Shoigu; the Main Department for Research and Technological Support of Advanced Technologies, itself part of the Russian MoD; and the Main Research and Testing Robotics Center of the Russian MoD, one of the most secretive military organizations in the country (Kozyulin 2019: 26). The brand new military innovation technopolis 'ERA' was built to contribute to the development and application of breakthrough technologies such as AI in the defence sector (Mil.ru 2020; Era-tehnopolis.ru n.d.). The technopolis will host laboratories and engineering units of Russia's leading research and industrial enterprises, as well as scientific companies of the MoD (Kozyulin 2019: 26). Military innovation will also be facilitated by Rostec (Edmonds et al. 2021: 34).

There have been a few high-profile civilian efforts in AI. For example, the United Instrument-Making Corporation announced the launch of a large-scale, AI-oriented research project in 2015. The project brought together more than 30 Russian companies, educational, and scientific organizations. Among other successful projects are Yandex, Russia's internet giant equivalent to Google, VisionLabs, a team of computer vision and machine learning experts who specialize in customer facial recognition for the banking sector and retail, and N-Tech.Lab, another group of experts in facial recognition technologies (Bendett 2018: 162).

Skolkovo Innovation Center, a tech campus on the outskirts of Moscow, was once established to foment 'start-up culture' and attract corporate venture capital (Dougherty and Jay 2018). Despite the growing diversity of the commercial sector, the number of start-ups is small compared to that in the US and China (Bendett 2018: 162; Edmonds et al. 2021: 34). According to some statistics, the number of AI start-ups in Russia can be measured in hundreds, while the same can be measured in thousands in the US (Markotkin and Chernenko 2020). These are, for the most part, large *state-owned* – or at least *state-controlled* – companies, rather than start-ups, that spearhead Russia's quest for AI (Edmonds et al. 2021: 34). What testifies to this is the formation of the so-called *AI Russia Alliance*. Its current members are SberBank, Gazprom Neft, Yandex, Mail.ru Group, MTS, and the Russian Direct Investment Fund. Their shared role is, inter alia, to facilitate and coordinate efforts of the business and scientific communities in the sphere of AI (Yastrebova 2019; Markotkin and Chernenko 2020).

The backbone of the US AI 'ecosystem' is its *bottom-up* innovation culture, in contrast to both China and Russia. Tech firms compete for market share and do not function as instruments of state power (NSCAI 2021: 25). AI is, consequently, being developed for the most part not by the government but by private companies such as Google and Microsoft (Özdemir 2019: 14). American society's 'garage start-up' mentality is another distinctive feature of the US approach to AI (NSCAI 2021: 25). According to Tracxn statistics, there are *over 10,000* AI start-ups in the US (Tracxn 2021). The US government does not have direct control over such companies which are, in their turn, not always willing to cooperate. What illustrates this well is the case of Project Maven, one of the Pentagon's projects intended to turn AI into a huge battlefield advantage. It was a collaborative project between the DoD and Google. The latter, however, decided to withdraw from the project due to increased normative pressure on tech companies and governments to keep meaningful human control over the selection and engagement of targets (Özdemir 2019: 22).

However, US government and military leadership have themselves taken active measures to enhance their AI R&D. The main responsible institution when it comes to the application of AI in national security and defence is the DoD. The DoD's unclassified investments in AI and autonomy have grown more than fourfold – from just over $600 million in 2016 to $2.5 billion in 2021 (Sayler 2020b: 2). DARPA, its advanced technology branch, has taken the lead in AI R&D. In 2018, the Pentagon also established the Joint Artificial Intelligence Center (JAIC), which has since assumed a crucial role in the transition from R&D to operational capabilities (Özdemir 2019: 16–17). The National Security Commission on Artificial Intelligence, an independent commission intended to develop recommendations for future directions in AI R&D to bolster US national security and defence, was also created with the assistance of the DoD (Sayler 2020b: 5; NSCAI n.d.).

There have also been efforts towards closer and more efficient collaboration between the government, industry, and academia. The White House established the Select Committee on Artificial Intelligence under the National Science and

Technology Council (NSTC) in 2018. This committee serves as the main advisor to the White House and is responsible for facilitating cooperation between the government, industry, and academia in matters related to AI R&D. The NSTC's Subcommittee on Machine Learning and Artificial Intelligence (MLAI) was in turn tasked with monitoring the development of AI within the government, in the private sector, and internationally, and consult on national research needs and priorities (Özdemir 2019: 14 and 16–17). In 2020, the US National Science Foundation (NSF) announced its intention to establish several AI research institutes, each focused on a different topic and led by a different university. These institutes will serve as national R&D hubs and bring together the government, industry, and academia (Kratsios and Liddell 2020; Savage 2020). To oversee and implement America's national AI strategy, the White House established the National Artificial Intelligence Initiative Office in 2021. The office serves as the central hub for coordination of efforts *across* the government, as well as *between* the government, the private sector, academia, and other stakeholders (Trumpwhitehouse.archives.gov n.d.). The Biden administration, building off efforts started under the Trump administration, also launched a new National AI Research Resource Task Force. It will bring together the nation's foremost experts from the government, private industry, and academia (Heckman 2021).

Having outlined the different AI 'ecosystems' that China, Russia, and the US have, it is important to highlight that such 'ecosystems' can also interact with each other in various ways. Chinese experts and government organizations are, at least to some extent, inspired by concepts and approaches developed in the US (Kania 2017). US government and think tank reports are translated, disseminated, and analysed on a regular basis in China (Allen 2019: 3). China has, according to some sources, even developed a drone-ship similar to the US Navy's Sea Hunter (Sutton 2020). The Chinese company Ziyan, which produces the attack helicopter drone Blowfish A2, has pursued international sales of the drone (Abadicio 2019). Russia, seeking beneficial partnerships in AI R&D, has entered into cooperative agreements with China and South Korea through Huawei and Samsung (Edmonds et al. 2021: iii). Therefore, such interactions can indeed take various forms and relate, among others, to norm formation, competitive emulation, technology transfer, and R&D. They can also be *un*intended, as the following example demonstrates. Much of China's top AI talent, in fact, ends up in the US. Chinese citizens were even involved in Project Maven as engineers working at Google (Mozur and Metz 2021). All these examples are indicative of continuous interaction between different organizational structures and, in particular, of what we conceptualized as an *interactional order* in Chapter 1.

New military practices, routines, and experiences

It is still too early to assess the transformative impact of AI on military routines, operations, and roles. However, in the selected examples below, we evaluate the possible range of the *ongoing* AI-RMA. First of all, AI will be adapted for all sorts of missions. For instance, AI is being incorporated into weapons systems, intelligence, surveillance and reconnaissance technologies, command and control systems, cyberspace and information operations, as well as logistics by the US DoD (Sayler 2020b: 10).

AI will penetrate all military branches. For example, there are indications that AI is being explored for different applications by the Chinese PLA Army (PLAA), Air Force (PLAAF), Navy (PLAN), Rocket Force (PLARF), and Strategic Support Force (PLASSF) (Ding 2018: 31–32; Kania 2020: 3). AI will be embedded into weapons and other military systems in all war-fighting domains. For instance, the US has been making genuine efforts towards the incorporation of AI into fighter aircraft and drones, ground vehicles, naval vessels, and space-based systems (Osborn 2019; Özdemir 2019: 17). Further and more specific examples were provided earlier in this chapter. AI technologies will, at the same time, help integrate capabilities *across* all military services and warfare domains. The US DoD's Joint All-Domain Command and Control (JADC2) framework is a perfect example. It is an AI-driven network that connects sensors from all military services and shares targeting data and other intelligence among forces on land, at sea, in the air, space, and cyberspace (Freedberg 2021). AI-enhanced systems will themselves form autonomous, multi-agent battlefield networks, as already tested and proven. China, for example, launched a record-breaking *swarm* of 119 drones, all equipped with intelligent software to communicate with each other, in 2017 (Özdemir 2019: 21).

New forms of combat training and simulations are being explored as well to increase combat preparedness. The Artificial Intelligence and War-Gaming National Finals were, for instance, organized by the Chinese Institute of Command and Control in 2017. In this competition intelligent machines held a decisive advantage over humans (Özdemir 2019: 21). In Russia, a new War Games Center was established at the General Staff Military Academy. It is outfitted with a hi-tech training apparatus, with one of the modes pitting human capabilities against AI. Another great example is the most modern robot targets developed for target practice in Russia. Each of them is built from a human mold and carries a dummy machine gun. Such robots are equipped with AI. They can orient themselves at the test site, take cover from incoming fire, and act in an organized group, thereby preventing human soldiers from getting used to a familiar target environment (Isaev 2020). Some intelligent military equipment has even been tested under battlefield conditions. A good example is the deployment of Russia's combat robot Uran-9 in Syria. Uran-9 is capable of performing certain tasks autonomously, which is, to a considerable degree, achieved thanks to AI (Degtyarev 2021; Knutov 2021; Ptichkin 2021). Wartime combat experience revealed a number of shortcomings in its operation and all of those have since then been eliminated by the developers (Tsygankova 2019). Another example is the use of Russia's kamikaze drone ZALA Lancet to destroy terrorists in Syria (Lenta.Ru 2021; Ria Novosti 2021b). ZALA Lancet can find and recognize targets because its software contains elements of AI (Aksenov 2019).

Chapter 1 considered the intended and unintended consequences of military transformations in other spheres of social life from a purely theoretical perspective. Now is the perfect moment to give an example of how this works. The prospect of intelligent machines taking over human roles in the *kill* chain has provoked a coordinated civil society response in the form of the Campaign to Stop Killer Robots. It is 'the latest in a series of transnational advocacy campaigns in the area of humanitarian disarmament' (Carpenter 2016: 58). It has united almost 200 non-governmental organizations (NGOs) based in various countries, gained broad public support, and successfully dragged into its agenda a good number of state governments, thousands of experts, over 20 Nobel Peace Prize laureates, as well as elements of the United Nations and the European Union (CSKR n.d.). The Convention on Certain Conventional Weapons (CCW) in Geneva has expectedly become the core global inter-governmental deliberation forum for addressing key challenges posed by emerging technologies in fully autonomous lethal weapons (Altmann and Sauer 2017: 132). The second part of the book focuses entirely on this particular aspect, or rather consequence, of the AI-RMA.

Concluding remarks

Our theoretically and empirically informed conclusion in this chapter is that the AI-RMA is *underway*. Even though lack of information prevents us from capturing all the subtle nuances, there is enough evidence in our claim. AI is exceptionally *versatile* and a game changer at all levels. It has a major impact on all military services and is being tailored for all war-fighting domains and a whole variety of military missions. New concepts and new organizational 'ecosystems' are being developed to accommodate change in a way that provides strategic advantages.

Weapons and other military systems, as well as dual-use systems that *can have* military applications are themselves being revolutionized. AI technologies open up a *mix* of different sorts of new machine capabilities. However, this does not signal a *clean* break with the past. AI is set to change the nature of the human-machine relationship but ideas about the human-machine nexus will not become redundant and obsolete. New functions are, to a considerable degree, being interfaced into tested and proven machine designs. The most straightforward examples, as also explained above, are Ripsaw M5 and Pantsir-SM. (The former is an upgraded version of the Ripsaw series of ground combat vehicles; the latter is a modernized version of the Pantsir family of anti-aircraft weapons.) This general finding testifies to the utility of our proposed concept and heuristic model of *composite systems* developed in Chapter 2.

At the same time, we linked these developments with a growing international movement against lethal autonomous weapons systems. The campaign itself, as well as the debates and political processes that have taken place against its background, are examined in the following chapters. However, this chapter contains relevant findings that provide the basis for the discussions to follow. It sheds light on the complex nature of AI and potential regulatory challenges. AI technologies are dual-use, which is a challenge on its own merits. Another challenge, as illustrated above, is a *switchable* mode that is becoming an increasingly important part in weapons design. This mode allows for the switch of control: a machine can be run in either autonomous mode or manual mode with human operators. AI can likewise be adapted for a wide range of functions on a large variety of devices. While humans remain in control in one form or another, especially when it comes to the actual use of force, it will be hard to determine how functions are distributed and whose decision to fire is *final*. AI-*assisted* weapons are, therefore, not synonymous with what we know as *weaponized* AI. This important difference is often neglected.

Notes

1 Our heuristic categorization of these systems is informed by the domains in which they are *based*. But we recognize that their *effects* can spill into other domains.

2 Putting AI in control of certain operations involving the use of nuclear weapons is one of the most sensitive aspects when it comes to the militarization of AI. It is for this reason, we assume, that it is a top-secret weapon and that there is a lack of evidence of an *explicit* connection between the torpedo's capabilities and AI.

References

ЯRobot (2018) 'Итоги Конференции "Искусственный Интеллект: Проблемы и Пути Решения"' ['Results of the Conference "Artificial Intelligence: Problems and Solutions"']. 22 March 2018. Available at: https://ya-r.ru/2018/03/22/itogi-konferentsii-iskusstvennyj-intellekt-problemy-i-puti-resheniya-14-15-marta-v-parke-patriot/.

Abadicio M (2019) 'Artificial Intelligence in the Chinese Military – Current Initiatives'. *Emerj*. Last updated 21 November 2019. Available at: https://emerj.com/ai-zsector-overviews/artificial-intelligence-china-military/.

Adamsky D (2018) 'From Moscow with Coercion: Russian Deterrence Theory and Strategic Culture'. *Journal of Strategic Studies* 41(1–2): 33–60.

AF.mil [Secretary of the Air Force Public Affairs] (2019) 'Air Force Releases 2019 Artificial Intelligence Strategy'. 12 September 2019. Available at: https://www.af.mil/News/Article-Display/Article/1959225/air-force-releases-2019-artificial-intelligence-strategy/.

AF.mil [Secretary of the Air Force Public Affairs] (2020) 'GatewayONE and AttritableONE Test Moves Joint Force One Step Closer to "IoT.mil", Demonstrates F-22, F-35 First Secure Bi-Directional Data Sharing'. 14 December 2020. Available at: https://www.af.mil/News/Article-Display/Article/2446122/gatewayone-and-attritableone-test-moves-joint-force-one-step-closer-to-iotmil-d/.

AFRL [Air Force Research Laboratory] (2021) 'AFRL Successfully Completes XQ-58A Valkyrie Flight and Payload Release Test'. 5 April 2021. Available at: https://www.afrl.af.mil/News/Article/2561387/afrl-successfully-completes-xq-58a-valkyrie-flight-and-payload-release-test/.

Aksenov P (2019), 'Экспонаты "Армии-2019": Дрон-Камикадзе, Робот-Экраноплан и Гражданский "Калашников"' ['Exhibits of "Army-2019": Drone-Kamikaze, Robot-*Ekranoplan* and Civilian "Kalashnikov"']. *BBC News*, 27 June 2019. Available at: https://www.bbc.com/russian/features-48778256.amp.

Aksenov P (2020) 'Пентагон Готовится к Учебным Боям Роботов с Летчиками-Истребителями' [The Pentagon Prepares for Robots' Training Battles with Fighter Pilots]. *BBC News*, 11 September 2020. Available at: https://www.bbc.com/russian/features-54106488.

Allen GC (2019) *Understanding China's AI Strategy: Clues to Chinese Strategic Thinking on Artificial Intelligence and National Security*. Center for a New American Security (CNAS), February 2019. Available at: https://s3.us-east-1.amazonaws.com/files.cnas.org/documents/CNAS-Understanding-Chinas-AI-Strategy-Gregory-C.-Allen-FINAL-2.15.19.pdf?mtime=20190215104041&focal=none.

Altmann J, Sauer F (2017) 'Autonomous Weapon Systems and Strategic Stability'. *Survival* 59(5): 117–142.

Anno T (2019) *National Identity and Great-Power Status in Russia and Japan: Non-Western Challengers to the Liberal International Order*. New York: Routledge.

Annual Report (2020) *American Artificial Intelligence Initiative: Year One Annual Report*. White House Office of Science and Technology Policy. February 2020. Available at: https://trumpwhitehouse.archives.gov/wp-content/uploads/2020/02/American-AI-Initiative-One-Year-Annual-Report.pdf.

Army Recognition (2014) 'Chinese-Made Sharp Claw 1 Mini UGV Unmanned Ground Vehicle Makes Its Debut AirShow China'. 18 November 2014. Available at: https://www.armyrecognition.com/airshow_china_2014_zhuhai_news_coverage_report_uk/chinese-made_sharp_claw_1_mini_ugv_unmanned_ground_vehicle_makes_its_debut_airshow_china_1811143.html.

Baranets V (2021) 'Подводный Ядерный Беспилотник "Посейдон": Что Известно о Российской "Торпеде Судного Дня"' ['Underwater Nuclear Drone "Poseidon": What Is Known about the Russian "Doomsday Torpedo"']. *Komsomolskaya Pravda*, 12 February 2021. Available at: https://www.kp.ru/daily/27239.5/4366945/.

Bendett S (2018) 'The Development of Artificial Intelligence in Russia'. In: Wright ND (ed) *AI, China, Russia, and the Global Order: Technological, Political, Global, and Creative Perspectives*. Strategic Multilayer Assessment (SMA). Available at: https://nsiteam.com/social/wp-content/uploads/2019/01/AI-China-Russia-Global-WP_FINAL_forcopying_Edited-EDITED.pdf.

Bisht IS (2021) 'Chinese "Airborne Aircraft Carrier" Launches Drone Swarm'. *The Defense Post*, 9 April 2021. Available at: https://www.thedefensepost.com/2021/04/09/china-aircraft-carrier-launches-drone-swarm/.

Boulanin V, Saalman L, Topychkanov P, Su F, Carlsson MP (2020) *Artificial Intelligence, Strategic Stability and Nuclear Risk*. Stockholm International Peace Research Institute (SIPRI), June 2020. Available at: https://www.sipri.org/sites/default/files/2020-06/artificial_intelligence_strategic_stability_and_nuclear_risk.pdf.

Brar A (2019) 'Analysis: How China Is Applying AI to Its Military'. *BBC Monitoring*, 21 May 2019. Available at: https://monitoring.bbc.co.uk/product/c200tsqy.

Burke EJ, Gunness K, Cooper CA, Cozad M (2020) *People's Liberation Army Operational Concepts*. Doc. No. RR-A394-1. RAND Corporation. Available at: https://www.rand.org/pubs/research_reports/RRA394-1.html.

Caffrey MB (2019) *On Wargaming How Wargames Have Shaped History and How They May Shape the Future*. Naval War College. Available at: https://digital-commons.usnwc.edu/cgi/viewcontent.cgi?article=1043&context=newport-papers.

Carpenter C (2016) 'Rethinking the Political/-Science-/Fiction Nexus: Global Policy Making and the Campaign to Stop Killer Robots'. *American Political Science Association* 14(1): 53–69.

Chan MK (2019) 'China and the U.S Are Fighting a Major Battle Over Killer Robots and the Future of AI'. *Time*, 13 September 2019. Available at: https://time.com/5673240/china-killer-robots-weapons/.

Chen S (2019) 'How China's Scavenger Satellites Are Being Used to Develop AI Weapons, Drones and Robots'. *Yahoo! News*, 22 April 2019. Available at: https://sg.news.yahoo.com/china-scavenger-satellites-being-used-105436059.html?guccounter=1.

Clark C (2019) 'US "Loyal Wingman" Takes Flight: AFRL & Kratos XQ-58A Valkyrie'. *Breaking Defense*, 7 March 2019. Available at: https://breakingdefense.com/2019/03/us-loyal-wingman-takes-flight-afrl-kratos-xq-58a-valkyrie/.

Crosby C (2020) 'Operationalizing Artificial Intelligence for Algorithmic Warfare'. *Military Review* (July-August): 43–51. Available at: https://www.armyupress.army.mil/Portals/7/military-review/Archives/English/JA-20/Crosby-Operationalizing-AI-1.pdf.

CSKR [Campaign to Stop Killer Robots] (n.d.) 'Membership List'. Accessed 2 September 2021. Available at: https://www.stopkillerrobots.org/members/.

Darpa.mil (2016) 'DARPA Director to Christen ACTUV Prototype Vessel'. 4 July 2016. Available at: https://www.darpa.mil/news-events/2016-04-07.

Davis M (2021) 'The Artificial Intelligence "Backseater" in Future Air Combat'. *The Strategist*, 5 February 2021. Available at: https://www.aspistrategist.org.au/the-artificial-intelligence-backseater-in-future-air-combat/.

Deamer K (2016) 'US Military's Robotic Submarine Hunter Completes First Tests at Sea'. *Live Science*, 4 August 2016. Available at: https://www.livescience.com/55662-darpa-submarine-hunter-completes-tests.html.

Defense World (2019) 'Unmanned Surface Vehicles'. 1 October 2019. Available at: https://www.defenseworld.net/video/108/Unmanned_surface_vehicles#.YYWMC9NKiu5.

Defense World (2020) 'Russian Pantsir Air Defence System Gets Artificial Intelligence Upgrade'. 28 July 2020. Available at: https://www.defenseworld.net/news/27519/Russian_Pantsir_Air_Defence_System_Gets_Artificial_Intelligence_Upgrade#.YYVQW9NKh0t.

Degtyarev A (2021) 'Воины Будущего: Какими Возможностями Обладают Поступившие на Вооружение ВС РФ Боевые Роботы "Уран-9"' ['Warriors of the Future: What Capabilities Do Combat Robots "Uran-9" Acquired by the Russian Armed Forces Possess']. *Zvezda*, 29 January 2021. Available at: https://tvzvezda.ru/news/2021129549-DxbAL.html.

Dempsey M (2020) 'Robot Tanks: On Patrol but Not Allowed to Shoot'. *BBC News*, 21 January 2020. Available at: https://www.bbc.com/news/business-50387954.

Ding J (2018) *Deciphering China's AI Dream: The Context, Components, Capabilities, and Consequences of China's Strategy to Lead the World in AI*. Future of Humanity Institute, University of Oxford, March 2018. Available at: http://www.fhi.ox.ac.uk/wp-content/uploads/Deciphering_Chinas_AI-Dream.pdf.

Domanska M (2019) 'The Myth of the Great Patriotic War as a Tool of the Kremlin's Great Power Policy'. *OSW Commentary* 316. Centre for Eastern Studies, 31 December 2019. Available at: https://www.osw.waw.pl/sites/default/files/Commentary_316.pdf.

Dougherty J, Jay M (2018) 'Russia Tries to Get Smart about Artificial Intelligence'. *The Wilson Quarterly*, Spring 2018. Available at: https://www.wilsonquarterly.com/quarterly/living-with-artificial-intelligence/russia-tries-to-get-smart-about-artificial-intelligence/.

Duncan PJS (2000) *Russian Messianism: Third Rome, Revolution, Communism and after*. London: Routledge.

Edmonds J, Bendett S, Fink A, Chesnut M, Gorenburg D, Kofman M, Stricklin K, Waller J (2021) *Artificial Intelligence and Autonomy in Russia*. Doc. No. DRM-2021-U-029303. CNA's Strategy, Policy, Plans, and Programs Division, May 2021. Available at: https://www.cna.org/CNA_files/centers/CNA/sppp/rsp/russia-ai/Russia-Artificial-Intelligence-Autonomy-Putin-Military.pdf.

Eitelhuber N (2009) 'The Russian Bear: Russian Strategic Culture and What It Implies for the West'. *Connections* 9(1): 1–28.

Era-tehnopolis.ru (n.d.) 'Первый Военный Инноград' ['The First Military Innovation-City']. Accessed 1 August 2021. Available at: https://www.era-tehnopolis.ru.

Ermarth FW (2006) *Russian Strategic Culture: Past, Present, and … In Transition?* Prepared for Defense Threat Reduction Agency, Advanced Systems and Concepts Office. 31 October 2006. Available at: https://irp.fas.org/agency/dod/dtra/russia.pdf.

Everstine BW (2020) 'U-2 Flies with Artificial Intelligence as Its Co-Pilot'. *Air Force Magazine*, 16 December 2020. Available at: https://www.airforcemag.com/u-2-flies-with-artificial-intelligence-as-its-co-pilot/.

Executive Order [Executive Order 13859 of 11 February 2019] (2019) *Maintaining American Leadership in Artificial Intelligence*. Federal Register 84(31). 14 February 2019. Available at: https://www.govinfo.gov/content/pkg/FR-2019-02-14/pdf/2019-02544.pdf.

Fedasiuk R (2020) *Chinese Perspectives on AI and Future Military Capabilities*. Policy Brief. Center for Security and Emerging Technology, August 2020. Available at: https://cset.georgetown.edu/wp-content/uploads/CSET-Chinese-Perspectives.pdf.

Freedberg SJ (2019) 'Textron Rolls Out Ripsaw Robot For RCV-Light … And RCV-Medium'. *Breaking Defense*, 14 October 2019. Available at: https://breakingdefense.com/2019/10/textron-rolls-out-ripsaw-robot-for-rcv-light-and-rcv-medium/.

Freedberg SJ (2021) 'Inside Russia's Robot Army: Rhetoric vs. Reality'. *Breaking Defense*, 30 April 2021. Available at: https://breakingdefense.com/2021/04/inside-russias-robot-army-rhetoric-vs-reality/.

Gentile G, Shurkin M, Evans AT, Grisé M, Hvizda M, Jensen R (2021) *A History of the Third Offset, 2014–2018*. Doc. RRA454-1. RAND Corporation. Available at: https://www.rand.org/pubs/research_reports/RRA454-1.html.

Ghose T (2016) '"Sea Hunter": World's First Unmanned Ship Stalks Subs'. *Live Science*, 8 April 2016. Available at: https://www.livescience.com/54354-darpa-launches-unmanned-submarine-hunter.html.

Glass P (2018) 'China's Robot Subs Will Lean Heavily on AI: Report'. *Defense One*, 23 July 2018. Available at: https://www.defenseone.com/technology/2018/07/chinas-robot-subs-will-lean-heavily-ai-report/149959/.

Godwin PHB (1992) 'Chinese Military Strategy Revised: Local and Limited War'. *The Annals of the American Academy of Political and Social Science* 519: 191–201.

Graham T (2019) 'Let Russia Be Russia: The Case for a More Pragmatic Approach to Moscow'. *Foreign Affairs* 98(6). Available at: https://www.foreignaffairs.com/articles/russia-fsu/2019-10-15/let-russia-be-russia.

Grishchenko N (2019) 'Войны Будущего: Как "Панцирь-СМ" Борется с Роботами' ['Future Wars: How "Pantsir-SM" Fights Robots']. *Rossiyskaya Gazeta*, 9 July 2019. Available at: https://rg.ru/2019/07/09/vojny-budushchego-kak-pancir-sm-boretsia-s-robotami.html.

Hambling D (2020) 'U.S. To Equip MQ-9 Reaper Drones with Artificial Intelligence'. *Forbes*, 11 December 2020. Available at: https://www.forbes.com/sites/davidhambling/2020/12/11/new-project-will-give-us-mq-9-reaper-drones-artificial-intelligence/?sh=563e78b87a8e.

Heckman J (2021) 'White House Partners with NSF to Stand Up National AI Research Resource Task Force'. *Federal News Network*, 15 June 2021. Available at: https://federalnewsnetwork.com/artificial-intelligence/2021/06/white-house-partners-with-nsf-to-stand-up-national-ai-research-resource-task-force/.

Helfrich E (2020) 'Satellites to Be Developed with AI to Enable Fast Software Upgrades'. *Military Embedded Systems*, 28 July 2020. Available at: https://militaryembedded.com/comms/satellites/satellites-to-be-developed-with-ai-to-enable-fast-software-upgrades.

IAV [Inside Autonomous Vehicles] 'Ripsaw M5 UGV Rips It Up at AUSA'. 16 October 2020. Available at: https://insideautonomousvehicles.com/ripsaw-m5-ugv-rips-it-up-at-ausa/.

Igumnova L (2011) 'Russia's Strategic Culture between American and European Worldviews'. *Journal of Slavic Military Studies* 24: 253–273.

India Times (2019) 'China Has Developed World's First Amphibious Drone Boat and It's Called Marine Lizard'. 15 April 2019. Available at: https://www.indiatimes.com/news/world/china-has-developed-world-s-first-amphibious-drone-boat-and-it-s-called-marine-lizard-365487.html.

Insinna V (2020a) 'These 3 Companies Will Build Prototypes for the Air Force's Skyborg Drone'. *Defense News*, 7 December 2020. Available at: https://www.defensenews.com/air/2020/12/07/these-three-companies-will-build-prototypes-for-the-air-forces-skyborg-drone/.

Insinna V (2020b) 'Here's Why the Valkyrie Drone Couldn't Translate between F-35 and F-22 Jets During a Recent Test'. *C4ISRNet*, 18 December 2020. Available at: https://www.c4isrnet.com/battlefield-tech/it-networks/2020/12/18/heres-why-the-valkyrie-drone-couldnt-translate-between-an-f-35-and-f-22-during-a-recent-test/.

Insinna V (2021) 'Valkyrie Drone Launches Even Smaller Drone from Inside Payload Bay'. *Defense News*, 5 April 2021. Available at: https://www.defensenews.com/air/2021/04/05/the-valkyrie-drone-launches-an-even-smaller-drone-from-inside-its-payload-bay/.

INSSG [Interim National Security Strategic Guidance] (2021) *Renewing America's Advantages*. White House, March 2021. Available at: https://nssarchive.us/wp-content/uploads/2021/03/2021_Interim.pdf.

Interfax (2019) 'Минобороны Показало Видеокадры Полигонного Испытания "Посейдона"' ['Ministry of Defence Shows Video Footage of "Poseidon's" Polygon Test']. 20 February 2019. Available at: https://www.interfax.ru/amp/651366.

Isaev K (2020) '"Восстание Машин": в Парке "Патриот" Прошли Первые Испытания Роботов-Мишеней' ['"Rise of the Machines: First Test of Target Robots Conducted in "Patriot" Park]. *Zvezda*, 19 December 2020. Available at: https://tvzvezda.ru/news/20201219844-UZq97.html.

Izvestia (2019) 'Испытания Новейшего Комплекса ПВО "Панцирь-СМ" Завершатся в 2021 Году' ['Tests of the Newest Air Defense Complex "Pantsir-SM" Will Be Completed in 2021']. 4 June 2019. Available at: https://iz.ru/885116/2019-06-04/ispytaniia-noveishego-kompleksa-pvo-pantcir-sm-zavershatsia-v-2021-godu.

Izvestia (2021) 'Способная Нести "Посейдоны" Атомная Подлодка Впервые Вышла в Море' ['Nuclear Submarine Capable of Carrying "Poseidons" First Out at Sea']. 26 June 2021. Available at: https://iz.ru/1184609/2021-06-26/atomnaia-podlodka-s-poseidonami-vpervye-vyshla-v-more.

Kania EB (2017) 'Quest for an AI Revolution in Warfare: The PLA's Trajectory from Informatized to "Intelligentized" Warfare'. *The Strategy Bridge*, 8 June 2017. Available at: https://thestrategybridge.org/the-bridge/2017/6/8/-chinas-quest-for-an-ai-revolution-in-warfare.

Kania EB (2018) 'Artificial Intelligence in Future Chinese Command Decision Making'. In: Wright ND (ed) *AI, China, Russia, and the Global Order: Technological, Political, Global, and Creative Perspectives*. Strategic Multilayer Assessment (SMA). Available at: https://nsiteam.com/social/wp-content/uploads/2019/01/AI-China-Russia-Global-WP_FINAL_forcopying_Edited-EDITED.pdf.

Kania EB (2020) *"AI Weapons" in Chinese Military Innovation*. Brookings & Center for Security and Emerging Technology, April 2020. Available at: https://www.brookings.edu/wp-content/uploads/2020/04/FP_20200427_ai_weapons_kania_v2.pdf.

Kania EB, Laskai L (2021) *Myths and Realities of China's Military-Civil Fusion Strategy, Technology & National Security Program*. Center for a New American Security, January

2021. Available at: https://s3.us-east-1.amazonaws.com/files.cnas.org/documents/Myths-and-Realities-of-China%E2%80%99s-Military-Civil-Fusion-Strategy_FINAL-min.pdf?mtime=20210127133521&focal=none.

Keck Z (2019) 'We've Got the Details on China's Submarine Drones'. *The National Interest*, 7 November 2019. Available at: https://nationalinterest.org/blog/buzz/weve-got-details-chinas-submarine-drones-94686.

Keller J (2019) 'Unmanned Battle Tank Could Serve as Robotic Wingman'. *Military & Aerospace Electronics*, 28 October 2019. Available at: https://www.militaryaerospace.com/unmanned/article/14069388/unmanned-battle-tank-robotic.

Khrolenko A (2020) 'Российский "Панцирь" Наносит Сокрушительный Удар по Американским Беспилотникам' ['Russian "Pantsir" Delivers a Crushing Strike Against American Drones']. *Sputnik News*. Published 1 October 2020. Updated 2 October 2020. Available at: https://uz.sputniknews.ru/20201001/15097790.html.

Knutov Y (2021) '"Уран-9", "Штурм" и Другие Интеллектуалы' ['"Uran-9", "Storm" and Other Intellectuals']. *AI News*, 21 June 2021. Available at: https://ai-news.ru/2021/06/uran_9_shturm_i_drugie_intellektualy.html.

Kokoshin AA (2006) Реальный Суверенитет в Современной Мирополитической Системе [*Real Sovereignty in Contemporary World Political System*]. ISBN 5–9739–0058-4.

Kokoshin AA, Baluevskii YN, Potapov VY (2015) 'Влияние Новейших Тенденций в Развитии Технологий и Средств Вооруженной Борьбы на Военное Искусство' ['The Impact of the Latest Trends in the Development of Technologies and Means of Armed Struggle on the Military Art'. Вестник Московского Университета [*Moscow University Bulletin*] 25(4): 3–22.

Kotkin S (2016) 'Russia's Perpetual Geopolitics: Putin Returns to the Historical Pattern'. *Foreign Affairs* 95(3). Available at: https://www.foreignaffairs.com/articles/ukraine/2016-04-18/russias-perpetual-geopolitics.

Kozyulin V (2019) 'A Russian Perspective'. Discussion Paper. In: *The Militarization of Artificial Intelligence*. Stanley Center for Peace and Security, UN Office for Disarmament Affairs (UNODA), and Stimson Center, August 2019. Available at: https://reliefweb.int/sites/reliefweb.int/files/resources/TheMilitarization-ArtificialIntelligence.pdf.

Kratsios M, Liddell C (2020) *The Trump Administration Is Investing $1 Billion in Research Institutes to Advance Industries of the Future*. Office of Science and Technology Policy, 26 August 2020. Available at: https://trumpwhitehouse.archives.gov/articles/trump-administration-investing-1-billion-research-institutes-advance-industries-future/.

Kreisher O (2019) 'DARPA Director Praises Navy's Aggressive Use of Autonomous Sea Hunter'. *Sea Power*, 1 May 2019. Available at: https://seapowermagazine.org/darpa-director-praises-sea-hunter/.

Kremlin.ru [Presidential Executive Office] (2019) 'Excerpts from Transcript of Meeting on the Development of Artificial Intelligence Technologies'. 30 May 2019. Available at: http://en.kremlin.ru/events/president/news/60630.

Kriner DL, Shen FX (2014) 'Reassessing American Casualty Sensitivity: The Mediating Influence of Inequality'. *The Journal of Conflict Resolution* 58(7): 1174–1201.

Kruglov A, Ramm A, Dmitriev E (2018) 'Средства ПВО Объединят Искусственным Интеллектом' ['Air Defense Systems Will Be Integrated with the Help of Artificial Intelligence']. *Izvestia*, 2 May 2018. Available at: https://iz.ru/733333/aleksandr-kruglov-aleksei-ramm-evgenii-dmitriev/sredstva-pvo-obediniat-iskusstvennym-intellektom.

Layton P (2018) *Algorithmic Warfare: Applying Artificial Intelligence to Warfighting*. Canberra: Air Power Development Centre.

Leidos (2019) 'An Inside Look at the Autonomous Vessels Changing Warfighting'. Sponsor Content. *Defense One*, 29 October 2019. Available at: https://www.defenseone.com/sponsors/2019/10/inside-look-autonomous-vessels-changing-warfighting/160768/.

Lenta.Ru (2019) '"Панцирь-СМ" Назвали Оружием Будущего' ['"Pantsir-SM" Called the Weapon of the Future']. 8 July 2019. Available at: https://lenta.ru/news/2019/07/08/pantsirsm/.

Lenta.Ru (2021) 'Удар Российского Дрона-Камикадзе по Боевикам в Идлибе Попал на Видео' ['Russian Drone-Kamikaze Strike against Militants in Idlib Caught on Video']. 17 April 2021. Available at: https://lenta.ru/news/2021/04/17/dron/.

Levin Y (2018) 'The American Context of Civil Society'. *Stanford Social Innovation Review*, 14 June 2018. Available at: https://ssir.org/articles/entry/the_american_context_of_civil_society#.

Lin J, Singer PW (2014) 'China's New Military Robots Pack More Robots Inside (Starcraft-Style)'. *Popular Science*, 11 November 2014. Available at: https://www.popsci.com/blog-network/eastern-arsenal/chinas-new-military-robots-pack-more-robots-inside-starcraft-style/.

Litovkin D (2020) 'Чем Пугает Америку Русский Атомный Беспилотник "Посейдон"' ['Why Russian Atomic Drone "Poseidon" Scares America']. *TASS Russian News Agency*, 20 November 2020. Available at: https://tass.ru/opinions/10057181.

Lockie A (2016) 'Trump Questions the US's Nuclear Arsenal: Here's How the US's Nukes Compare to Russia's'. *Business Insider*, 24 December 2016. Available at: https://www.businessinsider.com.au/trump-tweet-us-nuclear-weapons-vs-russia-2016-12.

Lorek L (2020) 'Hypergiant Partners with the U.S. Air Force to Develop the Chameleon Constellation, an Updatable Satellite System'. *Silicon Hills*, 30 June 2020. Available at: http://siliconhillsnews.com/2020/06/30/hypergiant-partners-with-the-u-s-air-force-to-develop-the-chameleon-constellation-an-updatable-satellite-system/.

Mahnken TG (2006) *United States Strategic Culture*. Prepared for Defense Threat Reduction Agency, Advanced Systems and Concepts Office. 13 November 2006. Available at: https://irp.fas.org/agency/dod/dtra/us.pdf.

Markotkin N, Chernenko E (2020) *Developing Artificial Intelligence in Russia: Objectives and Reality*. Carnegie Moscow Center, 5 August 2020. Available at: https://carnegiemoscow.org/commentary/82422.

Melnikov R (2019a) 'Российский Ударный Беспилотник "Охотник" Изменит Тактику Авиации' ['Russian Attack Drone "Hunter" Will Change Aviation Tactics']. *Rossiyskaya Gazeta*, 15 October 2019. Available at: https://rg.ru/2019/10/15/rossijskij-udarnyj-bespilotnik-ohotnik-izmenit-taktiku-aviacii.html.

Melnikov R (2019b) 'Ужасный Ущерб: США Оценили Результат Применения Ядерного "Посейдона"' ['Terrible Damage: US Assessed the Outcome of "Poseidon's" Nuclear Attack']. *Rossiyskaya Gazeta*, 19 January 2019. Available at: https://rg.ru/2019/01/19/uzhasnyj-ushcherb-ssha-ocenili-rezultat-primeneniia-iadernogo-posejdona.html.

Melnikov R (2019c) 'СМИ: Атомный "Посейдон" РФ Развивает под Водой Невероятную Скорость' ['Media: Russia's Atomic "Poseidon" Develops Incredible Speed Under Water']. *Rossiyskaya Gazeta*, 4 January 2019. Available at: https://rg.ru/amp/2019/01/04/smi-atomnyj-posejdon-rf-razvivaet-pod-vodoj-neveroiatnuiu-skorost.html.

Mil.ru [Russian Ministry of Defence] (2018) 'Конференция "Искусственный Интеллект: Проблемы и Пути Их Решения – 2018"' ['Conference "Artificial Intelligence: Problems and Ways to Solve Them – 2018"']. Available at: https://mil.ru/conferences/is-intellekt.htm.

Mil.ru [Russian Ministry of Defence] (2020) 'Военный Инновационный Технополис "ЭРА"' ['Military Innovation Technopolis "ERA"']. Available at: http://mil.ru/era/about.htm.

Mizokami K (2019a) 'F-35 Fighter Pilots Might Soon Fly with Robotic Wingmen'. *Popular Mechanics*, 23 May 2019. Available at: https://www.popularmechanics.com/military/aviation/a27560364/f-35-robotic-wingmen/.

Mizokami K (2019b) 'The Ripsaw M5 Could Become the Army's First Robo-Tank'. *Popular Mechanics*, 16 October 2019. Available at: https://www.popularmechanics.com/military/weapons/a29476669/ripsaw-m5/.

Mizokami K (2020) 'The Air Force's Secret New Fighter Jet Will Come with an R2-D2'. *Popular Mechanics*, 23 December 2020. Available at: https://www.popularmechanics.com/military/aviation/a35046713/air-force-secret-new-fighter-jet-ai-copilot-r2d2/.

Mizokami K (2021) 'The Air Force's AI Brain Just Flew for the First Time'. *Popular Mechanics*, 13 May 2021. Available at: https://www.popularmechanics.com/military/aviation/a36412460/air-force-ai-brain-first-flight-skyborg-details/.

Morgan FE, Boudreaux B, Lohn AJ, Ashby M, Curriden C, Klima K, Grossman D (2020) *Military Applications of Artificial Intelligence Ethical Concerns in an Uncertain World*. Doc. RR3139-1. RAND Corporation. Available at: https://www.rand.org/pubs/research_reports/RR3139-1.html.

Mozur P, Metz C (2021) 'A U.S. Secret Weapon in A.I.: Chinese Talent'. *The New York Times*. Published 9 June 2020. Updated 13 April 2021. Available at: https://www.nytimes.com/2020/06/09/technology/china-ai-research-education.html.

National Strategic Plan (2016) *The National Artificial Intelligence Research and Development Strategic Plan*. Networking and Information Technology Research and Development Subcommittee, National Science and Technology Council, Executive Office of the President of the United States, October 2016. Available at: https://www.nitrd.gov/pubs/national_ai_rd_strategic_plan.pdf.

National Strategic Plan (2019) *The National Artificial Intelligence Research and Development Strategic Plan: 2019 Update*. Select Committee on Artificial Intelligence, National Science and Technology Council, Executive Office of the President of the United States, June 2019. Available at: https://www.nitrd.gov/pubs/National-AI-RD-Strategy-2019.pdf.

National Strategy (2019) Национальная Стратегия Развития Искусственного Интеллекта на Период до *2030* [*National Strategy for the Development of Artificial Intelligence Over the Period Extending up to the Year 2030*]. Указ Президента РФ № 490 'О Развитии Искусственного Интеллекта в Российской Федерации' [Decree of the President of the Russian Federation № 490 'On the Development of Artificial Intelligence in the Russian Federation']. Trans. Available at: https://cset.georgetown.edu/research/decree-of-the-president-of-the-russian-federation-on-the-development-of-artificial-intelligence-in-the-russian-federation/.

Naval Technology (n.d.) 'Sea Hunter ASW Continuous Trail Unmanned Vessel (ACTUV)'. Accessed 3 September 2021. Available at: https://www.naval-technology.com/projects/sea-hunter-asw-continuous-trail-unmanned-vessel-actuv/.

Newdick T (2021) 'Stealthy Valkyrie Drone Uses Weapons Bay for First Time to Launch Smaller Drone'. *The Drive*, 5 April 2021. Available at: https://www.thedrive.com/the-war-zone/40068/xq-58a-valkyrie-uses-weapons-bay-for-first-time-to-launch-smaller-drone.

Next Generation Plan [New Generation Artificial Intelligence Development Plan] (2017) *Full Translation: China's 'New Generation Artificial Intelligence Development Plan'*. Trans. by Webster G, Creemers R, Triolo P, Kania EB. Available at: https://www.newamerica.org/cybersecurity-initiative/digichina/blog/full-translation-chinas-new-generation-artificial-intelligence-development-plan-2017/.

NSCAI [National Security Commission on Artificial Intelligence] (2021) *Final Report*. March 2021. Available at: https://www.nscai.gov/wp-content/uploads/2021/03/Full-Report-Digital-1.pdf.

NSCAI [National Security Commission on Artificial Intelligence] (n.d.) 'About'. Accessed 3 October 2021. Available at: https://www.nscai.gov/about/.

Nurkin T (2020) 'The Importance of Advancing Loyal Wingman Technology'. *Defense News*, 21 December 2020. Available at: https://www.defensenews.com/opinion/commentary/2020/12/21/the-importance-of-advancing-loyal-wingman-technology/.

OCIA [Office of Cyber and Infrastructure Analysis] (2017) *Narrative Analysis: Artificial Intelligence*. National Protection and Programs Directorate, US Department of Homeland Security, July 2017. Available at: https://info.publicintelligence.net/OCIA-ArtificialIntelligence.pdf.

Osborn K (2017) 'Air Force Chief Scientist Confirms F-35 Will Include Artificial Intelligence'. *Defense Systems*, 20 January 2017. Available at: https://defensesystems.com/articles/2017/01/20/f35.aspx.

Osborn K (2019) 'Pentagon Pursues AI for Space War to Stop Anti-Satellite Weapons'. *Warrior Maven*, 26 November 2019. Available at: https://warriormaven.com/air/pentagon-pursues-ai-for-space-war-to-stop-anti-satellite-weapons.

OSTP [Office of Science and Technology Policy] (2019) *Accelerating America's Leadership in Artificial Intelligence*. 11 February 2019. Available at: https://trumpwhitehouse.archives.gov/articles/accelerating-americas-leadership-in-artificial-intelligence/.

Özdemir GS (2019) *Artificial Intelligence Application in the Military: The Case of United States and China*. SETA Analysis, No. 51, June 2019. Available at: https://setav.org/en/assets/uploads/2019/06/51_AI_Military.pdf.

Pawlyk O (2021) 'Drone-Ception: Valkyrie Drone Launches Mini-Drone in New Air Force Test'. *Military News*, 7 April 2021. Available at: https://www.military.com/daily-news/2021/04/07/drone-ception-valkyrie-drone-launches-mini-drone-new-air-force-test.html.

Peck M (2019) 'Meet the Marine Lizard: Is China's New Tank All Hype?' *The National Interest*, 18 April 2019. Available at: https://nationalinterest.org/blog/buzz/meet-marine-lizard-chinas-new-tank-all-hype-53212.

Peck M (2020) 'Swarms of Chinese Helicopter Rocket Drones Could Overwhelm the U.S. Navy'. *The National Interest*, 21 August 2020. Available at: https://nationalinterest.org/blog/reboot/swarms-chinese-helicopter-rocket-drones-could-overwhelm-us-navy-167521.

Philipp J (2019) 'China Is Branding Anti-Satellite Weapons as "Scavenger Satellites"'. *The Epoch Times*. Published 5 May 2019. Updated 6 May 2019. Available at: https://www.theepochtimes.com/china-is-branding-anti-satellite-weapons-as-scavenger-satellites_2907825.html?welcomeuser=1.

Pifer S (2015) 'Russia's Perhaps-Not-Real Super Torpedo'. *Brookings*, 18 November 2015. Available at: https://www.brookings.edu/blog/order-from-chaos/2015/11/18/russias-perhaps-not-real-super-torpedo/.

Polyakova A (2018) 'Weapons of the Weak: Russia and AI-driven Asymmetric Warfare'. *Brookings*, 15 November 2018. Available at: https://www.brookings.edu/research/weapons-of-the-weak-russia-and-ai-driven-asymmetric-warfare/.

Presidential Decree [Decree of the President of the Russian Federation No. 642 of 1 December 2016] (2016) *On the Strategy for Scientific and Technological Development of the Russian Federation*. Available at: https://www.prlib.ru/en/node/680193.

Pressman A (2019) '"A.I., Captain": The Robotic Navy Ship of the Future'. *Fortune*, 22 May 2019. Available at: https://fortune.com/longform/leidos-sea-hunter-ai-navy-ship/.

Ptichkin S (2021) 'Искусственный Интеллект Становится в Строй: Зачем Нужны Боевые Роботы' ['Artificial Intelligence Becomes Operational: Why Fighting Robots Are Needed']. *Rossiyskaya Gazeta*, 24 May 2021. Available at: https://rg.ru/2021/05/24/iskusstvennyj-intellekt-stanovitsia-v-stroj-zachem-nuzhny-boevye-roboty.html.

Ramm A, Litovkin D, Andreev E (2017) 'В Войска Радиоэлектронной Борьбы Придет Искусственный Интеллект' ['Artificial Intelligence Will Complement the Electronic Warfare Troops']. *Izvestia*, 4 April 2017. Available at: https://iz.ru/news/675891.

Ramm A, Stepovoi B (2019) 'Один на Всех: ЗРК "Панцирь" Уничтожит Противника без Экипажа' ['One for All: Anti-Aircraft Missile System "Pantsir" Will Destroy the Enemy without a Crew']. *Izvestia*, 2 December 2019. Available at: https://iz.ru/948322/aleksei-ramm-bogdan-stepovoi/odin-na-vsekh-zrk-pantcir-unichtozhit-protivnika-bez-ekipazha.

Ramm A, Stepovoi B (2020) 'Видит Цель: "Былина" Сможет Атаковать Противника Без Участия Оператора' ['Sees the Target: "Bylina" Can Attack the Opponent Without Operator Participation']. *Izvestia*, 16 April 2020. Available at: https://iz.ru/1000101/aleksei-ramm-bogdan-stepovoi/vidit-tcel-bylina-smozhet-atakovat-protivnika-bez-uchastiia-operatora.

Randall I (2019) 'Robot TANK Named "Ripsaw M5" with Armour-Piercing Ammunition and On-board Drones is Built for the US Army'. *Daily Mail*, 29 October 2019. Available at: https://www.dailymail.co.uk/sciencetech/article-7625517/Ripsaw-M5-Robot-TANK-armour-piercing-ammunition-board-drones-built-Army.html.

Reuters (2018) 'Beijing to Build $2 Billion AI Research Park: Xinhua'. 3 January 2018. Available at: https://www.reuters.com/article/us-china-artificial-intelligence-idUSKBN1ES0B8.

Ria Novosti (2019) 'Последний Рубеж Обороны. Что Будет Сбивать Модернизированный "Панцирь-СМ"' ['The Last Line of Defense. What Will the Modernized "Pantsir-SM" Shoot Down']. Published 24 June 2019. Updated 2 December 2019. Available at: https://ria.ru/20190624/1555796206.html.

Ria Novosti (2020) 'В России Испытали Новейший Зенитный Комплекс "Панцирь-СМ"' ['The Newest Zen Complex "Pantsir-SM" Was Tested in Russia']. Published 6 April 2019. Updated 3 March 2020. Available at: https://ria.ru/20190406/1552448506.html.

Ria Novosti (2021a) 'Летчик-Испытатель Раскрыл Возможности Беспилотника С-70 "Охотник"' ['Fighter Pilot Revealed the Capabilities of the S-70 "Hunter" UAV']. Published 21 February 2021. Updated 22 February 2021. Available at: https://ria.ru/20210221/bespilotnik-1598515173.html.

Ria Novosti (2021b) 'Удар Российского Беспилотника по Боевикам в Сирии Попал на Видео' ['Russian Drone Attack on Militants in Syria Caught on Video']. 17 April 2021. Available at: https://ria.ru/20210417/udar-1728753228.html.

Roberts H, Cowls J, Morley J, Taddeo M, Wang V, Floridi L (2021) 'The Chinese Approach to Artificial Intelligence: An Analysis of Policy, Ethics, and Regulation'. *AI & Society* 36(3): 59–77.

Roper W (2020) 'Exclusive: AI Just Controlled a Military Plane for the First Time Ever'. *Popular Mechanics*, 16 December 2020. Available at: https://www.popularmechanics.com/military/aviation/a34978872/artificial-intelligence-controls-u2-spy-plane-air-force-exclusive/.

Sapolsky HM, Shapiro J (1996) 'Casualties, Technology, and America's Future Wars'. *US Army War College Quarterly: Parameters* 26(2): 119–127.

Savage N (2020) 'The Race to the Top among the World's Leaders in Artificial Intelligence'. *Nature*, 9 December 2020. Available at: https://www.nature.com/articles/d41586-020-03409-8.

Sayler KM (2020a) *Defense Primer: U.S. Policy on Lethal Autonomous Weapon Systems*. Doc. No. IF11150. Congressional Research Service (CRS). Updated 1 December 2020. Available at: https://crsreports.congress.gov/product/pdf/IF/IF11150.

Sayler KM (2020b) *Artificial Intelligence and National Security*. Doc. No. R45178. Congressional Research Service (CRS). Originally written by Daniel S. Hoadley. Updated 10 November 2020. Available at: https://sgp.fas.org/crs/natsec/R45178.pdf.

Seidel J (2019) 'Space Junk or Sabotage? Space Clean-Up Drones Could Have Military Implications'. *News.com.au*, 29 June 2019. Available at: https://www.news.com.au/technology/science/space/space-junk-or-sabotage-space-cleanup-drones-could-have-military-implications/news-story/e2cbbb479ea620a64a43ddc0d860c30c.

Sharapov A (2020) 'На Параде Победы Показали Нового "Убийцу" Дронов "Панцирь-СМ"' ['New Drone "Killer" "Pantsir-SM" Was Shown at the Victory Parade']. *Mk.ru*, 24 June 2020. Available at: https://www.mk.ru/politics/2020/06/24/na-parade-pobedy-pokazali-novogo-ubiycu-dronov-pancirsm.html.

Shield AI (2020) 'Textron Systems and Shield AI to Collaborate on Military Multi-Domain Autonomy Technologies'. Press Release, 13 October 2020. Available at: https://shield.ai/news/2020/10/13/textron-systems-and-shield-ai-collaborate-on-military-multi-domain-autonomy.

Slijper F, Beck A, Kayser D (2019) *State of AI: Artificial Intelligence, the Military and Increasingly Autonomous Weapons*. PAX for Peace, April 2019. Available at: https://paxforpeace.nl/media/download/state-of-artificial-intelligence--pax-report.pdf.

Stefanovich D (2020) 'Artificial Intelligence Advances in Russian Strategic Weapons'. Chapter 4. In: Topychkanov P (ed) *The Impact of Artificial Intelligence on Strategic Stability and Nuclear Risk: Volume III South Asian Perspectives*: 25–29. Stockholm International Peace Research Institute. Available at: http://www.jstor.org/stable/resrep24515.10.

Stepanov A (2019) 'Новейшие Комплексы ПВО "Панцирь-СМ" и С-350 Вскоре Поступят в Войска' ['The Newest "Pantsir-SM" and S-350 Air Defense Systems Will Soon Complement the Troops']. *Rossiyskaya Gazeta*, 19 June 2019. Available at: https://rg.ru/2019/06/19/novejshie-kompleksy-pvo-pancir-sm-i-s-350-vskore-postupiat-v-vojska.html.

Stewart P (2016) 'U.S. Military Christens Self-Driving "Sea Hunter" Warship'. *Reuters*, 8 April 2016. Available at: https://www.reuters.com/article/us-usa-military-robot-ship-idUSKCN0X42I4.

Strategy Page (2020) 'Armor: Chinese Combat Droids Evolving'. 17 May 2020. Available at: https://www.strategypage.com/htmw/htarm/20200517.aspx.

Strout N (2020) 'Hypergiant is Building a Reprogrammable Satellite Constellation with the Air Force'. *Defense News*, 27 July 2020. Available at: https://www.defensenews.com/battlefield-tech/space/2020/07/27/hypergiant-is-building-a-reprogrammable-satellite-constellation-with-the-air-force/.

Suciu P (2020) 'China's Army Now Has Killer Robots: Meet the "Sharp Claw"'. *The National Interest*, 17 April 2020. Available at: https://nationalinterest.org/blog/buzz/chinas-army-now-has-killer-robots-meet-sharp-claw-145302.

Summary of DoD Strategy [Summary of the 2018 Department of Defense Artificial Intelligence Strategy] (2018) *Harnessing AI to Advance Our Security and Prosperity*. 12 February 2019. Available at: https://media.defense.gov/2019/Feb/12/2002088963/-1/-1/1/SUMMARY-OF-DOD-AI-STRATEGY.PDF.

Summary of NDS [Summary of the 2018 National Defense Strategy of the United States of America] (2018) *Sharpening the American Military's Competitive Edge*. Available at: https://dod.defense.gov/Portals/1/Documents/pubs/2018-National-Defense-Strategy-Summary.pdf.

Sutton HI (2020) 'New Intelligence: Chinese Copy of US Navy's Sea Hunter USV'. *Naval News*, 25 September 2020. Available at: https://www.navalnews.com/naval-news/2020/09/new-intelligence-chinese-copy-of-us-navys-sea-hunter-usv/.
Tadjdeh Y (2019) 'AUSA NEWS: Textron Pitches Ripsaw for Army's Robotic Combat Vehicle Program'. *National Defense*, 15 October 2019. Available at: https://www.nationaldefensemagazine.org/articles/2019/10/15/textron-pitches-ripsaw-vehicle-for-armys-robotic-combat-vehicle-program.
Tadjdeh Y (2021) 'Algorithmic Warfare: Russia Expanding Fleet of AI-Enabled Weapons'. *National Defense*, 20 July 2021. Available at: https://www.nationaldefensemagazine.org/articles/2021/7/20/russia-expanding-fleet-of-ai-enabled-weapons.
Tang D (2019) 'China Expands Military Might with New Drone Boat'. *The Times*, April 15 2019. Available at: https://www.thetimes.co.uk/article/china-builds-world-first-armed-amphibious-drone-boat-vqwkkh5h0.
TASS [Russian News Agency] (2017) 'Путин: Лидер по Созданию Искусственного Интеллекта Станет Властелином Мира' ['Putin: The Leader in the Development of Artificial Intelligence Will Become the Ruler of the World']. 1 September 2017. Available at: https://tass.ru/obschestvo/4524746.
TASS [Russian News Agency] (2018a) 'Russia's Okhotnik Heavy Drone Makes First Ground Run'. 23 November 2018. Available at: https://tass.com/defense/1032118.
TASS [Russian News Agency] (2018b) 'Беспилотный Подводный Аппарат "Посейдон". Досье.' ['Unmanned Underwater Vehicle "Poseidon". Dossier.'] 19 July 2018. Available at: https://tass.ru/info/5388731.
TASS [Russian News Agency] (2020) 'Russia Laying Groundwork So Nobody Dares Fight Against It, Says Putin'. 2 March 2020. Available at: https://tass.com/politics/1125431.
Thomson B (2020) 'China Unveils "Small but Lethal" War Robot "Sharp Claw I" That's Armed with a Machine Gun and Night Vision'. *Daily Mail*, 16 June 2020. Available at: https://www.dailymail.co.uk/news/article-8426027/Chinas-war-robot-Sharp-Claw-armed-machine-gun-night-vision.html.
Tirpak JA (2020) 'Skyborg Drone Translates between F-35 and F-22 in Test'. *Air Force Magazine*, 16 December 2020. Available at: https://www.airforcemag.com/skyborg-drone-translates-between-f-35-and-f-22-in-test/.
Tracxn (2021) 'Artificial Intelligence Startups in United States'. Last updated 5 November 2021. Accessed 7 November 2021. Available at: https://tracxn.com/explore/Artificial-Intelligence-Startups-in-United-States.
Trevithick J (2019) 'Navy's Sea Hunter Drone Ship Has Sailed Autonomously to Hawaii and Back Amid Talk of New Roles'. *The Drive*, 4 February 2019. Available at: https://www.thedrive.com/the-war-zone/26319/usns-sea-hunter-drone-ship-has-sailed-autonomously-to-hawaii-and-back-amid-talk-of-new-roles.
Trumpwhitehouse.archives.gov (n.d.) 'Artificial Intelligence for the American People'. Accessed 14 October 2021. Available at: https://trumpwhitehouse.archives.gov/ai/.
Tsargrad (2018) 'Полный Автомат, Искусственный Интеллект: СУ-57 Доведут до Уровня Беспилотника Шестого Поколения' ['Full Automation, Artificial Intelligence: SU-57 to Be Brought to The Level of a Sixth Generation Drone']. 24 August 2018. Available at: https://tsargrad.tv/news/polnyj-avtomat-iskusstvennyj-intellekt-su-57-dovedut-do-urovnja-bespilotnika-shestogo-pokolenija_154201.
Tsygankova S (2019) 'Разработчики Устранили Недостатки Боевого Робота "Уран-9"' ['Developers Remedied the Shortcomings of "Uran-9" Fighting Robot']. *Rossiyskaya Gazeta*, 4 April 2019. Available at: https://rg.ru/2019/04/04/razrabotchiki-ustranili-vse-nedostatki-boevogo-robota-uran-9.html.

Tulnovskii E (2021) 'Военное Обозрение' ['Military Review']. *Zapolyarnaya Pravda*, 23 February 2021. Available at: https://gazetazp.ru/news/gorod/voennoe-obozrenie.html#.

USAF Annex (2019) *The United States Air Force Artificial Intelligence Annex to The Department of Defense Artificial Intelligence Strategy*. Available at: https://www.af.mil/Portals/1/documents/5/USAF-AI-Annex-to-DoD-AI-Strategy.pdf.

USCC [The US-China Economic and Security Review Commission] (2015) *Annual Report to Congress of the US-China Economic and Security Review Commission*. 114th Congress, First Session, November 2015. Available at: https://www.uscc.gov/sites/default/files/annual_reports/2015%20Annual%20Report%20to%20Congress.PDF#.

Valagin A (2018) 'Выбран Прототип Российского Истребителя Шестого Поколения' ['Prototype Russian Sixth Generation Fighter Selected']. *Rossiyskaya Gazeta*, 20 July 2018. Available at: https://rg.ru/2018/07/20/reg-sibfo/vybran-prototip-rossijskogo-istrebitelia-shestogo-pokoleniia.html.

Vavasseur X (2019) 'China's Marine Lizard Amphibious USV'. *Naval News*, 16 April 2019. Available at: https://www.navalnews.com/naval-news/2019/04/chinas-marine-lizard-amphibious-usv/.

Vecchio A (2019) 'China Operating the Scavenger Satellites: Only in Anti-Debris Function?' *Difesa Online*, 18 June 2019. Available at: https://en.difesaonline.it/mondo-militare/cina-operativi-i-satelliti-spazzini-solo-funzione-anti-detriti.

Walsh JI (2015) 'Political Accountability and Autonomous Weapons'. *Research and Politics* 2(4): 1–6.

Warrick J (2021) 'China Is Building More than 100 New Missile Silos in Its Western Desert, Analysts Say'. *The Washington Post*, 30 June 2021. Available at: https://www.washingtonpost.com/national-security/china-nuclear-missile-silos/2021/06/30/0fa8debc-d9c2-11eb-bb9e-70fda8c37057_story.html.

Wolfe F (2021) 'Sixth Flight Test of XQ-58A Valkyrie Features First Weapons Bay Release'. *Aviation Today*, 5 April 2021. Available at: https://www.aviationtoday.com/2021/04/05/sixth-flight-test-xq-58a-valkyrie-features-first-weapons-bay-release/.

Wood P (2019) 'Chinese Shipbuilder Launches Amphibious "Sea Iguana" Unmanned Surface Vehicle'. *Ashtree Analytics*, 22 May 2019. Originally available at: https://www.ashtreeanalytics.com/posts/chinese-shipbuilder-launches-amphibious-sea-iguana-unmanned-surface-vehicle. Accessed 15 October 2021.

Xuanzun L (2019a) 'Chinese Helicopter Drones Capable of Intelligent Swarm Attacks'. *Global Times*, 9 May 2019. Available at: https://www.globaltimes.cn/page/201905/1149168.shtml.

Xuanzun L (2019b) 'China Builds World's First Armed Amphibious Drone Boat that Can Lead Land Assault'. *Global Times*, 14 April 2019. Originally available at: https://www.globaltimes.cn/content/1145839.shtml. Currently available at: http://english.chinamil.com.cn/view/2019-04/15/content_9477847.htm.

Yastrebova S (2019) 'Чем Займется Российский Альянс по Развитию Искусственного Интеллекта' ['What the Russian Alliance for the Development of Artificial Intelligence Will Do']. *Vedomosti*, 9 November 2019. Available at: https://www.vedomosti.ru/technology/articles/2019/11/09/815838-alyans.

Zakrzewski C (2021) 'The Senate Approved a Massive Investment in U.S. Tech Competitiveness'. *The Washington Post*, 9 June 2021. Available at: https://www.washingtonpost.com/politics/2021/06/09/technology-202-senate-approved-massive-investment-us-tech-competitiveness/.

Part II

Autonomous Weapons Systems

Politics and Operations of Power

4 Dilemmas in Autonomous Weapons Systems

Introduction

The rise of artificial intelligence (AI) and the revolution in military affairs (RMA) debate which it re-energized have attracted a wave of public and media attention over the last decade. In the previous chapters we developed tools for analysing and understanding the AI-RMA. One of the most important findings is that we are facing a multifaceted phenomenon which cannot be reduced to a series of simple assumptions. However, rapidly increasing attention has been directed to one particular aspect oof this revolution: the delegation of *authority* and *capability* to initiate the use of lethal force to a machine (Asaro 2012: 688). Autonomous weapons are set to be the centre of attention and they are described as the world's third greatest revolution after gunpowder and nuclear arms (FLI 2015). The so-called *killer robots* have become one of the most popular cultural imaginaries of the twenty-first century. The Campaign to Stop Killer Robots has emerged as 'the latest in a series of transnational advocacy campaigns in the area of humanitarian disarmament' (Carpenter 2016: 58).

The debate on killer robots frequently leads to a grossly oversimplified and inaccurate interpretation, deliberate or inadvertent, of the AI-RMA. At the very least, it shifts attention from *military functions* of AI to as-yet imagined *weaponized* AI. The much broader category of *military* systems includes weapons, intelligence, surveillance, and reconnaissance tools, as well as vehicles used for the delivery of cargo, supplies, and even people. None of them is immune from the effects of AI, with the prospect of them all being transformed (McFarland 2015: 1329–1330). AI-*assisted* weapons are also clearly not the same thing as *weaponized* AI. The following section, inter alia, makes clear that the political construction of increasingly autonomous weapons *as* killer robots verges on science fiction rather than a precise representation of imminent reality. Due to the fact that much of the object of the debate exists in the non-tangible, imagined, and speculative realm, an analysis of pros and cons needs to be carried out in a careful and meticulous manner.

This chapter opens the second part of this book. Its starting point is to examine the nature of current weapons programmes against the background of how killer robots are being imagined and conceived. This is what allows us to go beyond the emotive, one-sided, and analytically unproductive term 'killer robots'.

DOI: 10.4324/9781003045489-7

Our academic preference and choice lie with a more neutral perspective and term: autonomous weapons systems (AWS). The reason we begin with this analysis is to show that AWS, not yet in existence, should be seen in connection with a broad range of possible applications for AI. In the first part of this book, the AI-RMA was discussed and analysed on a more general level. AI was conceptualized as a new *layer* of systems capabilities deposited onto the existing layers in *composite systems*. Attention was drawn to *evolutionary* characteristics of the AI-RMA. One of the main arguments was that technologies do not stop being *tools* in the hands of humans even though they are being endowed with more and more complex functions. The following section extends this discussion to AWS.

Having presented our understanding of AWS, we proceed to the intellectual reconstruction of the debate itself. We start anew given the manifold of speculative imagination around AWS. We do not seek to bring a normative perspective into the discussion and do our best to represent the debate in an impartial manner. The best available means for outlining both pros and cons in a balanced way is *dilemma analysis*. In particular, this chapter identifies seven dilemmas that need to be reconciled on the way towards some sort of regulatory response. Delimiting dilemmas, we draw attention to a dichotomy that underlies the debate: AWS are a mixture of good and bad.

In what follows below we set the baseline for the next two chapters which delve into the *politics* of AWS. One cannot make sense of the positions, interests, and identities of the involved actors without understanding the core of the debate. Only after differing views are presented and their importance appreciated will we move on to analyse the balance of political power between advocacy groups and blocking coalitions. The reflection on current and future technological realities presented here will also help us to reveal the *limits* of securitization practices of the Campaign to Stop Killer Robots.

Killer robots and the gap between imagined and existing technology

AWS are defined by the International Committee of the Red Cross as armed robots that 'can independently select and attack targets, i.e. with autonomy in the "critical functions" of acquiring, tracking, selecting and attacking targets' (ICRC 2014: 7). The United States Department of Defense defines them in the same way, in principle, as weapon systems 'that, once activated, can select and engage targets without further intervention by a human operator' (DoD Directive 2012/2017: 13). of particular concern are lethal autonomous weapon systems (e.g., UNODA 2017), lethal autonomous systems (e.g., Lucas 2010: 293), autonomous lethal technologies (e.g., Asaro 2012: 693), lethal autonomous weapons (e.g., Scharre 2018: Chap. 17), lethal autonomous robot weapons (e.g., Sharkey 2012: 790), fully autonomous weapons (e.g., HRW 2012), fully autonomous armed robots (e.g., Sharkey 2010: 370), and fully autonomous robotic weapons (e.g., O'Connell 2014: 526). The variety of these terms allows us to capture two distinctive and most controversial features of AWS: *full autonomy* and *lethality*. With these characteristics in mind, we can define AWS by their ability to (a) operate without human control and supervision in dynamic, unstructured, and open environments; (b) engage in autonomous (including lethal) decision-making, targeting, and force; (c) combine defensive and *offensive* attributes; (d) learn and adapt their behaviour (Sharkey 2010: 370, 2012: 787; Asaro 2012: 690; Kastan 2013: 49; FLI 2015; Altmann and Sauer 2017: 118). Complex systems of this kind are being made possible by advances in AI (O'Connell 2014: 526). AI algorithms will allow humans to become ever more detached from immediate decisions made and implemented on the battlefield (Walsh 2015: 2).

Policy advocates call for a *preventive*, or rather *preemptive*, global ban on the development, production, and use of AWS (FLI 2015; HRW 2016). Their view can be summarized as follows:

> A common misconception about lethal autonomous weapons is that they are far in the future and not a reality. The term 'killer robots' is excellent short form to clearly and concisely communicate what these systems are. It can also conjure up ideas of 'The Terminator,' a self-aware, anthropomorphized, conscious robot assassin, which can fuel the assumption that these weapons are science fiction. The reality is that weapons that can autonomously select, target, and kill humans are already here.
>
> (Stop Autonomous Weapons n.d.)

It is clarified elsewhere that they are *under development* (HRW 2018). The concept of a 'killer robot' has, as a result, penetrated ethical, legal, political, strategic, scientific, and academic discourses. It often relies on a horrifying representation of dangerous cinematic robots such as *The Terminator* (Carpenter 2016: 53). It also puts aside other uses of AI, especially those hidden from the public eye, and ignores the fact that the militarization of AI is a multifaceted, multidimensional, and far more complex process. Emotional appeals to the audience of non-experts and civilians would not succeed if complex technical language is used. The goal is to convey 'a simple and dramatic message' in layman's terms (Rosert and Sauer 2021: 21). Yet, at the same time, policy advocates seek to dispel the illusion that killer robots are the same as *Terminator*-like robots (Mary Wareham cited in Ghosh 2019; PAX n.d.). This leads to confusion among the public about how they differ and, as a result, causes misunderstandings, misrepresentations, and superficial speculations.

Nevertheless, the discourse on killer robots is about autonomous weapons of the future, whose time will perhaps never come, rather than about weapons that are currently operational and deployable. That said, we must differentiate between autonomous weapons that are being imagined and sophisticated, partially, and fully automated weapons systems that now exist. The so-called *unmanned* combat vehicles (e.g., MQ-1 Predator, MQ-9 Reaper, Talon SWORDS) can navigate autonomously but their selection and engagement of targets requires human input (Lele 2017: 59). Israel's Harpy is one of the most advanced unmanned weapons considered to be a precursor to AWS (HRW 2012). It is a suicide drone which can select a target based on radar signals and engage with it thereafter (Horowitz 2016: 91). It can loiter for hours before detecting, locking onto, and destroying its target. However, it is designed for use against hostile radars rather than against humans (Finn and Scheding 2010: 178). What is more, the parameters of its autonomous mode are pre-determined and humans decide on the target area (Vallor 2016: 212; Brenneke 2018: 65). Humans can also insist upon target verification (Finn and Scheding 2010: 178). There have been no major changes to the upgrades of the Harpy (the Harop and the Harpy NG). Both of them follow pre-programmed rules, serving as no more than anti-radiation missiles in their fully autonomous mode and reserve the right for a human to exercise control over task performance (Boulanin and Verbruggen 2017: 54).

Weapons systems that are able to identify, track, and attack targets on their own also exist. These are counter-rocket, artillery, and mortar systems (e.g., Iron Dome), missile defence systems (e.g., Patriot and Aegis), anti-aircraft systems (e.g., S-400), and close-in weapons systems (e.g., the Phalanx CIWS). Lele (2017: 59) characterized them as 'the only fully autonomous weapon systems that are completely operational'. However, most of them are sort of 'electric fences' (Johnson and Axinn 2013: 137–138). It is because such weapon platforms are stationary, fixed in their parameters, and designed for defence against *inanimate* targets (Altmann and Sauer 2017: 118). Humans decide where to deploy them, when to activate their so-called autonomous 'mode', and can override their operation at any time (Walsh 2015: 2; Horowitz 2016: 89–90). The most significant of all the precursors to AWS is the SGR-A1 (PAX n.d.). It is a stationary weapon platform designed to guard the demilitarized zone between North and South Korea. This system is distinct in that it classifies *human beings* detected in this zone as targets. However, it is placed in a structured environment to which human access is 'categorically prohibited' (Tamburrini 2016: 126). Even though its architecture supports the autonomous 'mode', humans supervise its operation via camera links (Wakefield 2018).

Roff (2015) was right in arguing that software which guides most of the existing weapons is 'relatively "stupid" from an AI perspective'. More sophisticated weapons systems are being developed owing to further advancements in AI. The greatest public attention and fear is focused upon the prospect that killer robots will '*eliminate* [emphasis added] human judgement in the initiation of lethal force' (Asaro 2012: 693). However, no weapons system to date has been designed in a way that it can *decide* to engage human targets *on its own*. AI will assist, rather than *displace*, the human decision maker when it comes to the deployment of lethal force. Intelligent functions have, as we argue, come to form the next *layer* in the multilayered structure of capabilities possessed by weapons systems. Our theoretical grasp of what we call *composite systems* is presented in Chapter 2. Examples are provided in Chapter 3.

Here we particularly focus on breakthroughs in systems capabilities enabled by AI. The goal is to understand whether *fully* autonomous *lethal* weapons, aka killer robots, are imminent. To the authors' knowledge, six intelligent functions are being developed in next-generation weapons: (1) enhanced situational awareness;

(2) decision support; (3) target acquisition; (4) interoperability; (5) recognition of human-selected targets; and (6) lethal decisions in controlled circumstances.

The first can be broken down into two distinct advancements: AI will improve situational awareness for both machines themselves and for humans. Both can be illustrated with real-world examples. The US NF-16 VISTA has proved in a series of flight tests that drones will be able to detect and adapt to unforeseen obstacles *with no human intervention* (Rosenberg 2017). Russia's Marker, a robotic platform used for testing the most advanced technologies for combat modules, is being trained to understand human-to-robot voice commands (TASS 2020). The US F-35's advanced sensor fusion algorithms will, in turn, acquire, distill, and organize otherwise disparate pieces of intelligence into a single integrated picture *for the pilot* (Osborn 2017). Another example is the US MQ-9 Reaper. Its sensing capabilities intended to assist the human operator will be enhanced with object recognition algorithms based on AI (Defense Post 2020).

The second function is included in the list because intelligent algorithms and smart sensors will augment human capabilities in more than one way. They will collect, process, and structure various types of information to be used in *human decision-making*. For instance, China's PLAN nuclear submarines will be equipped with decision support systems of this sort to enhance commanding officers' thinking skills and reduce their workload and mental burden (Chen 2018; Kania 2018). The third function on the list offers a narrower perspective on the same application of AI. It concerns targeting decisions in particular. The US Advanced Targeting and Lethality Automated System (ATLAS) will, for instance, use machine learning algorithms to *help* humans spot threats, prioritize targets, and bring the gun to bear but will not pull the trigger itself (Freedberg 2019). Russia's T-14 Armata will also be equipped with a fire control system that will search for targets, recognize, prioritize, and track them (Ria Novosti 2021). However, its so-called *autonomous* turret will be remote controlled by the crew seated in the isolated capsule *inside* the tank (Army Technology n.d.). Even though this battle tank was tested in the unmanned mode, it is not intended for serial production in this configuration (Ramm 2021).

The fourth intelligent function listed above will manifest in at least two different forms: systems integration and human-machine nexus. The Russian Aerospace Forces have, for instance, tested an automated control system with elements of AI. It will combine air defence systems and early warning radars into a single 'armoured fist', perform realtime situation analysis, and issue recommendations for the use of weapons (Kruglov, Ramm, and Dmitriev 2018). The same can be illustrated by the US Navy's 'Super Swarm'. This project has focused on the development of massive swarms of drones, unmanned submarines, and unmanned surface vessels which will co-ordinate their attacks against ships. The first successful test of a swarm attack in which a swarm of drones destroyed a surface vessel has been completed (Hambling 2021). Closer integration of human soldiers and machines will also be observed. An excellent example is Russia's Nerekhta-2. It will derive targeting data from the weapon of the human soldier *synchronized* with it and will assist in destroying the targets. It will be programmed and trained to understand human commands and gestures as well (Valagin 2017; Stanavov 2020).

The fifth on the list of intelligent functions under development is exemplified by the US Long Range Anti-Ship Missile (LRASM). Its mission will be to arrive at the target area and use its sensors and on-board software 'to *confirm* [emphasis added] the human-selected target' (Scharre 2018: Chap. 4). Another similar example is the US Cannon-Delivered Area Effects Munition (C-DAEM). It will search an area to identify and attack moving vehicles, in particular enemy tanks and heavily clad shells, far from human oversight. It will not be able to *choose* its *own* target but will find and engage the target identified by the human observer around its reported position (Dialani 2019; Lye 2020).

The development of the sixth intelligent function can be best illustrated by the autonomous fire control module presented by the Kalashnikov Group and equipped with AI. Keller (2017) called it 'a real life *Terminator*'. First of all, this module will be able to recognize, illuminate, and track targets and will be compatible with all combat modules produced by the Kalashnikov Concern. Whenever installed onboard, it will be a valuable asset to the human operator. However, it can also be switched to an autonomous mode in which it will scan the operational space, detect hostile objects, distinguish between *persons* and machines, determine priorities in the sequence of defeat, decide on the required number of shots for guaranteed destruction of each, and *open fire*. If unarmed, and therefore harmless, people – or civilians – appear in the operational space, the module will steer the fire aside. Since artificial neural networks modelled on the human brain will be used to structure its software, the module will be able to *learn* in the process of operation. Kalashnikov Media, a media platform that reports on the whole variety of products and services of the Kalashnikov Group, has released a video that presents the module's autonomous mode behaviour. The video makes it clear that this mode will be *activated* by a human and our interpretation of this fact is that its deployment will be a tactical decision (Kalashnikov Media 2018; TASS 2018a). Andrey Koshkin, an expert in military and political affairs, opined that this fire control module will be most useful in case of the enemy's active fire and for the protection of military facilities. He also assumed it might be incorporated into the architecture of the Uran-9 (Ria Fan 2018). On the subject of the Uran-9, TASS military observer Viktor Litovkin clarified, however, that such combat modules are themselves not *autonomous* to the extent believed (TASS 2018b).

The line between weapons capabilities being developed at the moment and imagined killer robots is thin, to say the least, but none of the mentioned systems 'have the power to *determine when* [emphasis added] to take human life' (HRW 2014). Therefore, it is perhaps premature to speak of *preemption* as long as the threat is less immediate than it is often claimed to be. What remains to be considered is *preventive* prohibition. This option, howbeit, comes with its own risk: a simple ban will miss the complexity of the issue (Chatham House 2018). It is noteworthy, for instance, that parallels are drawn between the case of laser weapons and that of AWS (HRW 2015). However, the Protocol on Blinding Laser Weapons (1995) did in fact target a *specific* weapon with a *well-defined harmful effect*: permanent blindness (Sivakumaran 2012: 399). The same cannot be said about AWS. Some argue the delegation of life and death decisions to machines is unethical, immoral, and should be made illegal (Asaro 2012: 708). This argument takes it for granted that human conduct is ethical ('humans are ethical, and robots are not') (Zawieska 2017: 49). Others insist that, on the contrary, it is a 'moral imperative' to make use of AWS (Arkin 2018: 321). This is an implied criticism of human ethics and morals ('humans fail to act ethically, so we need ethical robots') (Zawieska 2017: 49).

Whether the costs of their deployment exceed the benefits, or vice versa, has become a subject of debate, especially in the light of the existing trends and problems. The following sections provide insight into the kinds of arguments and counter-arguments that have been raised and taken into account. The goal is not to take sides, question one's arguments, or draw a line between the opponents and proponents of AWS. It is to *structure* the debate itself. Seven *dilemmas*, each revealing two sides of the same coin, are identified.

Dilemma No. 1: Predictability and controllability

The first dilemma associated with fielding fully autonomous weapons concerns the question of whether humans will be able to understand and predict the outcomes of their performance. This dilemma is placed at the forefront because it is of fundamental importance and underlies multiple issues related to the other dilemmas. Two sets of arguments have emerged in the discussion of how predictable and controllable these weapons *can* and *should* be.

On one hand, it is increasingly difficult to predict the behaviour of complex autonomous systems (Sparrow 2007: 65). Humans might lose control over – and perhaps even knowledge of – what is being done in their name (Sparrow 2009a: 25). The problem cannot be resolved by determining the parameters of respective computer programs. No magic formula or algorithm can replicate all the nuances and qualities that constitute human intelligence in a machine (Klincewicz 2015: 167). The debate is that of software *rigidity* versus *flexibility*. With rigid instructions to govern them, robots can misbehave if events and situations unforeseen or insufficiently imagined by the programmers occur. It is impossible to code them for all eventualities and all military scenarios accordingly. What is more, environments in which they are tested differ from more complex, unstructured, and dynamic battlefield conditions (Lin, Bekey, and Abney 2008: 20, 40, and 78). Their decisions will therefore be based on whatever information is available at the time they are programmed (McFarland 2015: 1335). If they face situations not readily fitting their designed parameters, they might fail due to the lack of situational awareness and common sense reasoning (Scharre 2011: 92; Asaro 2012: 691; Sharkey 2012: 789). There is an alternative: equipped with machine learning software, robots will act by rules that are not fixed during the production process and can be changed by the robots themselves in the process of their operation (Matthias 2004: 177). They will independently interpret, manage, and categorize large volumes of data, act upon that information, and learn from their experience

(Ayoub and Payne 2016: 793–794). However, it might be too hard to predict with a reasonable degree of certainty how they will behave in new situations and new contexts and what they will learn (Lin, Bekey, and Abney 2008: 8 and 40). There is a risk that being able to overwrite the contents of their own software, fully autonomous weapons will not follow any strict, predetermined rules, including the laws of war and rules of engagement (Lin, Bekey, and Abney 2008: 66; Ayoub and Payne 2016: 816). The most radical of these fears is that robots will one day run amok (Lin, Bekey, and Abney 2008: 78).

Even if software is designed in a way that strikes a balance between rigidity and flexibility, its *inherent weaknesses* are beyond remedy. Each computer program has bugs, i.e., errors in the logic of the program itself which can manifest only during its execution (Klincewicz 2015: 168). Nor are they immune to interferences (Noone and Noone 2015: 33). Bugs can be exploited by hackers to cause programs to do something other than what they are designed for on a regular basis (Klincewicz 2015: 169–170). Computer programs can also experience breakdowns, malfunctions, and glitches (Noone and Noone 2015: 33). Their increasing complexity is not without compromise: it creates conditions in which errors are more likely to happen. Different parts of a large, complex computer program can sometimes interact in unexpected, untested ways which makes it difficult to predict the effect of a given command with absolute certainty (Lin, Bekey, and Abney 2008: 8 and 79). The growing sophistication of such programs makes them even more sensitive and susceptible to cyber attacks (Klincewicz 2015: 163). Software failures can lead to accidents and fatal mistakes on the battlefield (Lin, Bekey, and Abney 2008: 79; Klincewicz 2015: 168). Cyber attacks could be employed to take control of autonomous, computer-controlled weapons and direct them against friendly forces or civilian populations (Schmitt 2013: 7). These vulnerabilities are magnified by the autonomous character of next generation weapons and the complex environment of the modern battlefield. Behaviour displayed by a large number of autonomous systems operating in concert to perform a designated military mission – i.e., so-called *swarm behaviour* – can lead to far more unpredictable outcomes (Kastan 2013: 52–53). The result of their autonomous interaction with hostile devices controlled by unknown software will be impossible to predict (Sharkey 2017: 182).

All of this becomes particularly disturbing in light of human limitations exposed by advances in AI. AWS will be able to process information and reach decisions at much faster speeds than humans (Ayoub and Payne 2016: 799). While humans might need a minimum of hundreds of milliseconds to make decisions, next generation weapons will make them in nanoseconds and that is at *superhuman speed* (Sharkey 2008a: 16; Gubrud 2014: 33; Scharre 2017: 26). If the speed of action on the battlefield is beyond the speed of human decision-making, the potential for events to spiral out of human control is obvious (Gubrud 2014: 39; Sharkey 2017: 182).

On the other hand, there is nothing new in these arguments. The problem of malfunction is not unique to AWS. It has proved pertinent to the variety of obsolete and existing weapons ranging from catapults to more complex computer-controlled combat systems (Schmitt 2013: 7; McFarland 2015: 1338). Threats posed by cyber attacks are not new either. The behaviour of human soldiers themselves can be unpredictable since they are subject to a number of psychological factors while engaged in armed conflict (Klincewicz 2015: 164 and 172). If the argument of the salience of these biases is accepted, AWS is perhaps not something to be resisted but, on the contrary, a perfect solution. Taking humans out of the decision-making chain, they promise to remove much of the uncertainty in human behaviour from the battlespace (Noone and Noone 2015: 30). They will also decouple the limits of weapon systems from that of their operators (Sparrow 2009a: 26).

At the same time, humans do not necessarily lose control over the combat outcomes when tasks are delegated to AWS. One point missed by this argument is that there is no such thing as *complete* autonomy so it is mistaken to imagine *independent* machines (McFarland 2015: 1323 and 1326–1327). Even systems endowed with the greatest degree of autonomy will never be 'human-free' (Schmitt 2013: 4). It might only appear that a given system can *choose* between alternatives; in fact, the choice is made in advance by the programmers. This choice is an exercise in software development, and it does not matter whether rigid instructions are programmed into a machine or whether it is designed to learn and adapt its behaviour intelligently (McFarland 2015: 1327–1328 and 1335). Since a machine's capacity for making decisions remains more or less fixed, its *decisions* are in either case a matter of time lag between a human decision and its anticipated combat effect (Robillard 2018: 711). That said, AWS should not be considered as detrimental to the existing command and control structure. Human combatants are expected to act in accordance with the laws of war, humanitarian provisions, treaty obligations, and specific rules of engagement. Full autonomy has seldom been granted even to human commanders. Nor will it be granted to AWS (Lucas 2014: 323–324).

Dilemma No. 2: Further dehumanization of killing

Another dilemma lies in the replacing of human soldiers with intelligent machines on the battlefield. The use of drones has already become associated with the *dehumanization* of killing (Wagner 2014: 1410). AWS go a step further in the dehumanization of warfare (Sharkey 2012: 799). The physical distance from the act of killing will not be greater. The only difference is that the psychological, mental, and moral distance between a human operator and a killing machice will no longer play a significant role either (Wagner 2014: 1411).

On one hand, this might mean the transformation of the process of killing into an 'unempathic automated industrial process' (Korać 2018: 62). AWS will have no emotions (Johnson and Axinn 2013: 129). Human emotions, in particular *healthy* ones, do play a positive role in combat (HRW 2012; Johnson and Axinn 2013: 136). Among them are the innate reluctance to kill, the ability to empathize and feel compassion, guilt, concern, and mercy (Krishnan 2009: 130;

Lucas 2014: 327; Wagner 2014: 1410 and 1415; Birnbacher 2016: 120). These stand as major obstacles to using lethal force, especially against civilians, in war (HRW 2012, 2016; Korać 2018: 51). AWS will be capable of *soulless* killings (Scharre 2018: Chap. 17). Not only will it be easier for them to kill but they will also deprive enemy soldiers and non-combatants of even hope for empathy, mercy, and reprieve (Birnbacher 2016: 121; HRW 2016).

It is also difficult to imagine all the circumstances in which soldiers might find themselves. The choice of the appropriate course of action is in many situations up to their best judgement (Lin, Bekey, and Abney 2008: 32). Human judgement and reason are necessary to make decisions in accordance with the laws of war whose formulations are often imperfect, incomplete, and subject to interpretation. At the moment of writing, human situational awareness exceeds the capacity of computer software (Asaro 2012: 700). The fact that the distinction between civilians and combatants is poorly defined and cannot be reliably programmed is of particular concern. It is legally permissible to kill enemy combatants but not innocent civilians. However, combatants cannot be killed unnecessarily and do retain the right to surrender. Wounded and mentally ill soldiers also receive immunity. There are also situations in which it is legally permissible to kill civilians (Asaro 2008: 60–61). The distinction is therefore 'not just a matter of uniform' (Sharkey 2008b: 86). It is even more blurred in guerrilla and insurgent warfare in which combatants pose as civilians (Asaro 2008: 60). It may be a challenge for robot sensors to distinguish between a civilian carrying a weapon and a combatant or between a civilian carrying a walking stick and a combatant carrying an AK-47 (Kastan 2013: 48 and 60). AWS will most likely lack situational awareness and will not be able to apply common sense to assist in discriminating decisions (Sharkey 2012: 789). No doubt the result might be 'a hecatomb of innocent victims' (Birnbacher 2016: 118). What is more, the principles of practical law do not necessarily exhaust those of fundamental morality (Asaro 2008: 60). The ability to think and behave morally is another distinctive human feature. Moral reasoning cannot be codified or programmed either (Johnson and Axinn 2013: 135; Robillard 2018: 705). AWS compromise on all kinds of direct human involvement and thus will not be adequately informed of what types of force are legitimate and morally appropriate (Asaro 2012: 688). They will not be able, unlike humans, to purposefully disobey illegal or immoral orders either (Lin, Bekey, and Abney 2008: 65; Johnson and Axinn 2013: 135).

On the other hand, such an oversimplified idealization of human warfare can result in misleading conclusions. The 'mere potentiality' of a soldier's mercy, empathy, and compassion does not necessarily materialize on the battlefield (Birnbacher 2016: 121). Wars have been waged and fought for as long as history is recorded so they form an integral part of human nature (Arkin 2010: 334). On top of that, humans themselves have a 'dismal record' in ethical behaviour on the battlefield (Arkin 2018: 318). While some emotions can restrain humans for good cause, others can unleash their instincts and cloud their judgement (Arkin 2010: 333; Schmitt 2013: 13). The latter include negative emotions, emotional distortions, and other psychological factors such as frustration, fear, stress, hysteria, panic, spite, hatred, anger, hate, prejudice, revenge, vengefulness, resentment, mental disturbance, and the instinct of self-preservation (Lucas 2014: 326; Klincewicz 2015: 164; Noone and Noone 2015: 29–30; Birnbacher 2016: 119; Arkin 2018: 318). The lack of offensive spirit can under certain circumstances have a negative effect on the outcomes of military campaigns and the duration of war (Arkin 2010: 337).

Biological limitations can also negatively affect human performance on the battlefield. These include the requirement of breathable air, rest and sleep, drinkable water and food, the physical extremes of acceleration and cognitive load, as well as vulnerability to temperatures, radiation, and biological and chemical weapons (Gubrud 2014: 38). Last but certainly not least, military decisions are made and actions are taken amidst the 'fog of war' (Arkin 2010: 333; Noone and Noone 2015: 32; Korać 2018: 56). In the military context, interactions are carried out in noisy and stressful conditions. They are also challenged by time pressures, environmental hazards, and degraded communications. Some decisions have to be made based on unclear orders and contradictory information (Arkin 2010: 333 and 336; Scharre 2011: 92; Ayoub and Payne 2016: 798).

It is therefore possible that humans underreact or overreact in certain cases and certain conditions, whereby underreaction is as damaging as overreaction (Noone and Noone 2015: 32). The former unnecessarily prolongs the horror and misery of war (Scharre 2018: Chap. 17). The latter results in excessive and indiscriminate uses of force, war crimes, friendly fire incidents, and unjustified collateral damage (Lucas 2014: 326; Klincewicz 2015: 164; Arkin 2018: 319). We are driven back again to the same point: AWS are perhaps the saviours, not the destroyers, of the lives of mankind. From a solely technical perspective, AWS programmed to comply with particular laws will have no other choice (Scharre 2018: Chap. 17). Even though they will be able to optimize their algorithms based on self-evaluation and learning, the problem of their *freedom* to choose whether or not comply with these laws can be solved by technological 'fixes' (Sharkey 2012: 791). They will be devoid of emotional content, which inter alia means their behaviour on the battlefield will not be influenced by *negative* emotions (Altmann and Sauer 2017: 119). They will also be resilient to psychological and biological performance-hindering conditions that sometimes underlie the perpetration of unlawful acts by humans (Lin, Bekey, and Abney 2008: 1; Liu 2012: 649). More still, they can at least partially offset the adverse effects of the 'fog of war' (Sparrow 2009a: 26). This is because they will render vulnerable control and communication links between human operators and remotely operated platforms functionally obsolete (Altmann and Sauer 2017: 119). Their own decisions will be based exclusively on bias-free, data-driven analysis (Ayoub and Payne 2016: 799).

AWS do have the potential to save civilians and combatants from being killed unlawfully or accidentally (Noone and Noone 2015: 25). If programmed accordingly, they will put an end to conscious and deliberate violations of rules of war (Lucas 2010: 293). Doing away with moments of hesitation and mercy when killings are *objectively* necessary to end wars sooner, they will save more human lives overall (Scharre 2018: Chap. 17). AWS might therefore, and quite paradoxically, make war *less inhumane* by eliminating the human element (Noone and Noone 2015: 29). If they are made illegal, in certain cases it might mean the denial of legal protection to soldiers and civilians (Wagner 2014: 1420).

Dilemma No. 3: Continuous depersonalization of the enemy

AWS are also associated with the increasing depersonalization of war (Klincewicz 2015: 163). What that means in practical terms is that the enemy is being more and more depersonalized, that is objectified (Sharkey 2012: 788; Korać 2018: 54). At the same time, new and future weapons systems are designed to increase the personalization in the deployment of force (Heyns 2016: 7). This presents us with still another dilemma.

On one hand, the use of lethal force has traditionally been viewed as an interpersonal affair: a human is expected to take this decision and be physically present at the moment of its execution (Heyns 2016: 3). This practice has recently undergone some revision and killing at a distance has become more common (Sparrow 2009a: 26; Jha 2016: Chap. 5). There are consequences for the perception of the enemy, eventually reduced to the status of a mere set of 'targets' in a dislocated reality (Korać 2018: 59). AWS authorized to use lethal force without direct human oversight will take it to the next level (Heyns 2016: 4). While contemporary warfare is increasingly characterized by distance and detachment (e.g., drone warfare), AWS will exacerbate this trend and encourage *moral* disengagement from the use of deadly force (Sharkey 2010: 369; Garcia 2015: 61). Situational decisions of human soldiers and operators will be replaced by more general choices made by the programmers even before they are made on the battlefield (McFarland 2015: 1335 and 1339). These are algorithms that will make the final decision on whether or not to kill (Heyns 2016: 11; HRW 2016; Scharre 2018: Chap. 17). Warfare will therefore turn into what Sharkey (2012: 788) characterized as 'a factory of death'. Whoever is the enemy will become even less visible, less tangible, less animate, and will thus be deprived of human attributes and moral value (Korać 2018: 49 and 60). The still greater psychological distance might also make it far easier for humans to kill *in absentia*, that is by means of the deployment of AWS (Sparrow 2009a: 26).

On the other hand, AWS do not in themselves imply a fundamentally different quality of warfare, altogether new and distinct from historical and contemporary experiences (Birnbacher 2016: 121). Much of modern warfare *is*, in fact, mechanical slaughter and impersonal killing at a distance (Scharre 2018: Chap. 17). While the former is apparent and does not need comment, the latter is evident, for instance, with the use of improvised explosive devices, over-the-horizon and indirect-fire weapons (Noone and Noone 2015: 33). So, there is enough evidence to argue that it does not matter in the end, provided that all the other parameters of the situation are equal, whether the victim is killed by a manned or unmanned weapon (Birnbacher 2016: 120). It is because there is no principled difference between being mowed down by a machine gun, suffocating from a sucking chest wound, whatever its source, or being blasted to bits by a bomb (Scharre 2018:

Chap. 17). Noone and Noone (2015: 33) also drew attention to the fact that 'seeing the man's eyes as he stabs you doesn't make your death any more palatable'. If biases are put aside, one might realize that the deployment of fully autonomous weapons could have positive effects in practice. AWS are expected to be more precise and accurate in their targeting than most, if not all, existing weapons and can therefore save lives and prevent unwarranted injuries (Wagner 2014: 1411; Birnbacher 2016: 119). It may sound paradoxical, but is probably true, that the increased depersonalization in the deployment of force might, as a matter of fact, mean highly *personalized* outcomes (Heyns 2016: 7).

Dilemma No. 4: Coordinated operations and the nexus between humans and AWS

The next dilemma refers to the scenario in which lethal autonomous weapons are *integrated* into wars fought between humans (Schmitt 2013: 6). AWS are not expected, at least initially, to entirely replace human soldiers on the battlefield (Heyns 2013: 9). It is far more probable that they will be deployed alongside fellow troops in some sort of concerted efforts (Wagner 2014: 1419).

On one hand, the creation of armed forces composed of human warfighters and intelligent machines, themselves delegated with lethal authority, puts human beings 'in harm's way' (Sparrow 2009b: 172). The most pronounced fear is to fall into the illusion of a *push button* war which can be fought at a distance and without casualties (Asaro 2008: 62; Sparrow 2009a: 26). AWS will perhaps be trusted too much and fielded with the intention that they can complete the full array of military tasks with no immediate risks to one's combatants and civilians. However, their capability to produce decisive victory in whatever circumstances might turn out to be very limited. If AWS fail to complete the tasks they are assigned, humans will have to take the baton, placing their own lives at risk. To protect valuable intelligence, fellow combatants will even have to defend, service, and repair them behind enemy lines (Sparrow 2009b: 173). With the tendency to anthropomorphize intelligent machines, human soldiers also run the risk of treating them as fellow warriors and being prepared to sacrifice themselves *nobly* for them (Sharkey 2012: 792).

There are other challenges too. The distinction between friends and enemies is often value-laden. There is no readily available algorithm, at least for the moment, that will determine a unique and clearcut set of criteria for each (Asaro 2008: 61). AWS will therefore be unable to discriminate between them with certainty and there is a high probability of friendly fire incidents (Kastan 2013: 53). It is also possible that the enemy will resort to cyber means in order to take control of one's autonomous weapon and direct it against one's friendly forces or civilian population (Schmitt 2013: 7). If equipped with video cameras and sensors to record and report whatever takes place on the battlefield, AWS will perhaps erode trust among fellow soldiers and negatively affect the team's cohesion and performance as well (Lin, Bekey, and Abney 2008: 79–80). They will also be ill-equipped to conceive dynamic changes in goals, tactics, and strategies which are often a necessary response to the demands of today's complex, ambiguous, and unstable operational environments (Ayoub and Payne 2016: 797). If a conflict between their software and real combat demands arises, intelligent and autonomous battlefield systems can refuse to follow an otherwise legitimate order (Lin, Bekey, and Abney 2008: 74).

On the other hand, AWS will be used to keep human beings 'out of harm's way' (Heyns 2016: 6). The exposure of fellow soldiers to dangerous, and life-threatening missions is being minimized as much as it can be (Arkin 2017: 37). *Unmanned* weapons have already proved to be extremely effective because they are developed, deployed, and operated by humans with 'no skin in the game' (Johnson and Axinn 2013: 133). Perhaps no other means can be as militarily effective and politically practical as them if the goal is to engage and defeat adversaries with a minimum of friendly casualties (Lin, Bekey, and Abney 2008: 54; Arkin 2018: 318). AWS will go far beyond ensuring the physical well-being of those who are on active duty: they will mitigate the most serious psychological effects of war as well. By eliminating the need for human operators wherever possible, they will avoid the traps of cognitive overload and protect at least some military personnel from the lasting mental effects of traumatic war experiences (Arkin 2010: 336–337; Noone and Noone 2015: 33).

Other biases are also easy to shed. The argument that friendly fire incidents will be a harmful consequence of the deployment of *killing* machines does not hold in the real world. History tells us that such incidents have often been the result of human error of some kind (Noone and Noone 2015: 32). Such errors can be the result of emotional disturbance and battle fatigue on the part of human combatants. Here it is worth highlighting that there are situations in which they deliberately decide to act outside of the chain of command, sometimes with wide-ranging consequences (Klincewicz 2015: 164). Unlike humans, who can deviate from orders, AWS will do what they are programmed to, so the commanders will in fact have greater control over how their forces behave on the battlefield (Scharre 2017: 22). AWS would, by default, be more considerate of the safety of human beings and give hope for fewer injuries and deaths by friendly fire (Klincewicz 2015: 164). If programmed to monitor the battlefield and report whatever ethical infractions are observed, they might also reduce the opportunities for unethical conduct by fellow comrades (Arkin 2010: 333, 2018: 321).

Dilemma No. 5: AWS and wider political and strategic considerations

The deployment of fully autonomous weapons also poses a major strategic dilemma. On one hand, human strategy is hard to measure. This is because it accompanies the instrumental use of violence in the pursuit of non-quantifiable, usually social, goals. It is subject to change in response to emerging situations and opportunities, and the effects of cultural and psychological processes can also be influential. AWS will be limited in their ability to grasp, internalize, and reproduce the finer nuances of human strategy (Ayoub and Payne 2016: 796–797).

If they are permitted to decide independently at different – strategic, operational, tactical – levels of war, humans risk losing control over the strategy formulation process, the conduct of military operations, and even entire wars in terms of their initiation, escalation, and termination (Ayoub and Payne 2016: 814; Scharre 2017: 22; Korać 2018: 57–58).

AWS represent a far more difficult challenge at the structural level. They are 'expendable' and can be risked in provocative military ventures (Gubrud 2014: 39). This conviction can create the illusion of a *risk free* war (Sharkey 2008a: 16). Political gains are always measured against acceptable costs of war, in particular wartime casualties. If political calculus does not have to take into account the number of fallen soldiers, the internal legitimization of military engagements abroad will be much easier. Civilian populations will be bribed by the fact that machines, not friends and relatives, will be sent to war (Wagner 2014: 1419–1420). They might eventually become disengaged from such decisions, no longer being associated with a huge physical and emotional toll, and defined almost exclusively in economic and diplomatic terms (Heyns 2013: 11). Popular resistance to military aggression will be removed as a result (Liu 2012: 633; Gubrud 2014: 38). The threshold for the use of force will also be lowered and armed conflict will cease to be the exceptional measure of last resort (Asaro 2012: 692; Sharkey 2017: 182). The effect of lowering the threshold will be the *normalization* of war (Heyns 2013: 11). Conflicts occurring below the threshold and low intensity wars will be completely out of sight (Korać 2018: 61). The army of machines is nothing but an illusion, however, because they will only supplement, not entirely replace, human combat forces (Kastan 2013: 53).

As soon as one nation is capable of deploying AWS, other nations will have a powerful incentive to catch up (Sparrow 2007: 69). Then two major scenarios are equally possible: this arms race might become a truly *global* one, and these weapons become the 'the Kalashnikovs of tomorrow'. It is because AWS require no costly or hard to obtain raw materials, especially as most of the enabling technologies are *dual-use* (Gubrud 2014: 39; FLI 2015; Altmann and Sauer 2017: 132). They will become a new class of weapons of mass destruction, while being much cheaper, much easier to build, and much more scalable than nuclear weapons (Stuart Russell cited in NYT 2019). Their proliferation will occur via exports, including to the grey and black markets (Altmann and Sauer 2017: 127). The higher priority will in many cases be given to the speed of proliferation rather than effective and technically feasible safeguards (Canning 2002: 32). It will only be a matter of time before rogue states, terrorists, criminal cartels, and extremist groups acquire AWS (Heyns 2013: 18; FLI 2015; Altmann and Sauer 2017: 127).

The imagination of a different, but equally frightening, scenario is inspired by the realization that cutting edge weapons are often 'the product of a rich and elaborate economy'. So will be AWS (Asaro 2008: 63). If this argument holds true, their development and deployment will create yet another imbalance of power in the international system: that of one between the haves and the have-nots (Garcia 2018: 339). In practical terms, it indicates the growing prominence of *asymmetric warfare*. Killing machines will often be pitted against human beings (Heyns

2013: 13 and 16). In certain cases, the encounter will even take the form of what Heyns (2013: 12) referred to as *one-sided killing.* The danger will not be less imminent for those who deploy AWS. Perhaps the best available means of retaliation and reprisals by the other side will be terrorism, including deliberate attacks on the enemy's civilian population (Asaro 2008: 63; Sharkey 2008a: 16; Heyns 2013: 16). Peaceful reconciliation will also be very difficult to achieve in cases of conflict between human beings and soulless machines (Lin, Bekey, and Abney 2008: 80).

AWS will also increase the risk of *accidental war* (Sparrow 2009a: 27). Herein lies another systemic challenge. Since they will make decisions at inhuman speeds, the danger is that unintended conflicts or unwanted escalation will be triggered outside of human intent and control (Asaro 2012: 692; Sharkey 2017: 183). It should also be expected that the amount of time available to the other side to detect an incoming attack and decide how to respond will decrease (Sparrow 2009a: 26). Miscalculations and misinterpretations, which can invite needless preemptive action and undesired escalation, will be more than probable under these conditions (Gubrud 2014: 39). The world's armed forces are therefore under pressure to mobilize and this further increases the probability that war can start by mistake (Sparrow 2009a: 27). The likelihood of accidental war has one implicit consequence with two sides: accidents can be attributed to the decisions of intelligent machines and the very possibility of it happening can *shield* human perpetrators from the responsibility for their own wartime behaviour (Asaro 2012: 693).

On the other hand, strategy that involves humans, in whichever capacity they act, will inevitably retain its 'human flavour'. AWS will be able to improve the quality of *human* strategic decisions but will not engage in independent priority setting and strategic planning (Ayoub and Payne 2016: 807–808 and 816). Their military utility at the tactical and operational levels cannot be denied either (Liu 2012: 633). AWS will extend the distance of the warfighter's reach and the depth and width of the battlefield due to their ability to see and strike farther (Arkin 2017: 37). They will also be better informed than the best human could be and will react much faster (Liu 2012: 633; Kastan 2013: 52; Lucas 2014: 325; Arkin 2018: 318). It is because they will absorb and process enormous amounts of data more efficiently, more effectively, and at a much higher speed (Wagner 2014: 1413; Korać 2018: 57). They will also be capable of learning and dynamic adaptation, unlike humans, who are often rooted in their entrenched preconceptions, cognitive biases, groupthink, and instinctive behaviour (Ayoub and Payne 2016: 807–808).

The problem of timelags in remote control also slows down the speed of human decision-making in combat situations (Heyns 2013: 10). Control and communication links between human operators and remotely controlled weapons are also vulnerable to electronic countermeasures, environmental factors, and other exigencies of the 'fog of war' (Sparrow 2007: 68). AWS will render such links obsolete (Altmann and Sauer 2017: 119). Their ability to operate with no human intervention and oversight is an obvious military advantage (Sparrow 2009a: 27). Their potential lies in force multiplication as well. Not only will each of them accomplish many of the functions now performed by humans but they will also allow fewer personnel to do more (Lin, Bekey, and Abney 2008: 1; Arkin 2017: 36, 2018: 318). AWS will be equipped with swarm intelligence algorithms which will allow them to imitate the hive mind behaviour of ants, bees, wasps, and other social insects (Dehuri, Cho, and Ghosh 2011: 1–2). It means one human operator will be able to initiate a large-scale attack by a swarm of robots (Sharkey 2008b: 87).

There are more ambitious expectations too. Endowed with greater autonomy, precision, and accuracy, AWS will render war less destructive and less indiscriminate (Lucas 2014: 326). The futuristic prospect is that humans will fight with each other only *by means of* AWS. If it materializes, 'war will cease to be a desirable option by nation-states as a means of resolving their differences', in some cases at least, as long as it carries no existential risk and boils down to inflicting economic damage (Lin, Bekey, and Abney 2008: 54). AWS can even assist humanity in transcending the *causes* of armed hostilities, both material and cultural, because they will not be able to comprehend their meaning. They give humans a chance for a world with fewer wars (Leveringhaus 2016: 7).

Dilemma No. 6: Applicability of existing legal frameworks to AWS

The debate is also conditioned in its substance and content by the question of whether international law is to be applicable to AWS. On one hand, AWS will flourish and operate in the so-called 'lawless zone' (Kastan 2013: 47). Making it easier to launch wars, they might erode respect for the principles of *jus ad bellum* (Lin, Bekey, and Abney 2008: 88). International law has evolved so as to constrain the use of force and make armed conflict the option of last resort (Heyns 2013: 11). The Charter of the United Nations, one of the key documents of international law, provides for two exceptions to this prohibition on the use of force: in exercising the right to self-defence and to solve problems that pose a threat to international peace and security. AWS will render this norm more fragile (Garcia 2015: 60). Nor will they be able to properly discriminate between combatants and civilians, as required by *jus in bello* (Lin, Bekey, and Abney 2008: 88). This poses a challenge to international humanitarian law (IHL) and international human rights law (IHRL) (HRW 2016).

The principle of human *dignity* and the right to *life* lie at the heart of IHRL (HRW 2014; Garcia 2015: 60–61). In the conditions that prevail during armed conflict, IHRL remains valid but is interpreted with reference to the rules of IHL (Heyns 2016: 8–10). IHL is based on two key principles and these are the principles of *distinction* and *proportionality* (McFarland 2015: 1337). They reflect the tension between two opposite goals: the need to differentiate legitimate military targets and the requirement that attacks against civilians and civilian objects must not be excessive given the military advantage expected (Wagner 2014: 1384–1385). AWS will be restricted in their abilities to interpret context and make value-based calculations (Heyns 2013: 11). So, the challenge is how to program them in such a way that these principles are kept intact (Korać 2018: 57). It is a challenge because humanitarian law contains contradictory and vague imperatives, suffers from terminological hurdles, and even lacks a clear definition of the civilian (Lin, Bekey, and Abney 2008: 76; Kastan 2013: 60). Subjective and context-specific estimates of value are often necessary to ensure compliance with all the applicable requirements (Heyns 2013: 13). AWS will undermine respect for the value of human life and the dignity of every human being, and therefore weaken the role of IHL (HRW 2014).

Article 36 of the Additional Protocol I to the Geneva Conventions also contains the obligation to conduct the *legal review* of new weapons before they are produced or acquired. This provision is to ensure that one's armed forces are equipped in accordance with one's obligations under international law. AWS will make it more difficult to detect non-compliance. This procedure demands a very high level of confidence that, once activated, they – as much as any other weapon – will operate as intended, both predictably and reliably (Davison 2017: 9–10). However, AWS and remotely operated weapons will appear identical from the outside. Their software can then be changed relatively quickly after the inspection, which will make cheating all too easy (Altmann and Sauer 2017: 135).

These legal gaps become of critical concern also because of the increasing *depersonalization* of responsibility (Heyns 2016: 12). What keeps the legal system on track is that individuals and states are held accountable for violations of legal rights and principles (Heyns 2013: 14). The attribution of responsibility for accidents and war crimes will be much vaguer in the age of AWS (Wagner 2014: 1371; Garcia 2018: 339), opening another legal 'gap' (e.g., HRW 2016). AWS will have either no one or too many people to be held responsible for their mistakes (Garcia 2015: 60). First of all, it is difficult to imagine that a machine itself should – and, in fact, could – be liable for damages it causes. Punishment is most effective when it entails suffering and is morally compelling, yet neither applies to AWS (Sparrow 2007: 71–72). There will also be no deterrent effect because one robot will not be deterred by the punishment of another robot (Kastan 2013: 68). Their designers, manufacturers, supervisors, and field commanders can be held accountable instead. So too can the procurement officers, government officials, and even presidents of respective countries (Lin, Bekey, and Abney 2008: 73). There is a risk, however, that military personnel and state employees will be charged for damage caused by the actions of machines which they could barely control at best (Sparrow 2007: 71).

The legal dimension of shared responsibility is also too weak to withstand the complexity of today's warfare environment. The existing approach to legal responsibility is 'atomized'. It seeks to link individual perpetrators with specific created effects. This approach runs contrary to the very idea of networks and swarms (Liu 2012: 650). The distribution of responsibility throughout complex networks, whose human and non-human, lethal and non-lethal elements are often inseparable and whose effects are often shared, will be arbitrary and unprincipled (McFarland 2015: 1333).

On the other hand, the lawfulness of weapons must be judged on a 'case-by-case basis' (Schmitt 2013: 8). It is not the weapons but the uses to which they are put that might be considered illegal (Birnbacher 2016: 118–121). Machine intelligence will not *automatically* cause superfluous injury or unnecessary suffering. What matters is the effect on the targeted individual, not the manner of engagement (Schmitt 2013: 9). Discrimination between soldiers and civilians, as well as reasoned proportionality judgements are 'no less difficult in air strikes and long-range attacks than they are with AWS' (Birnbacher 2016: 119). Even lawful weapons can be used unlawfully and humans themselves have a long history of systematic violations of international law (Schmitt 2013: 14). The 'fog of war' makes it more difficult to get all parties to comply with the laws of war (Korać 2018: 56). Potentially more accurate in distinction and proportionality calculations, AWS will perhaps outperform humans in ensuring respect for IHL (Arkin 2018: 323).

Also flawed is the argument that no one will be responsible for the actions of AWS. No machine is – or will ever be – autonomous to the extent that it is completely independent of 'human authorship'. Intelligent machines will be at the service of humans, which brings to the fore the principle of *human* responsibility (Birnbacher 2016: 120). The only difference will be that humans will be held accountable for the *deployment* of defective weapons, not for each selected target (Scharre 2018: Chap. 17).

Dilemma No. 7: The impact of culture and cultural diversity on the perception and regulation of AWS

The last, though certainly not least, dilemma concerns cultural aspects associated with the deployment of AWS. First of all, increasingly autonomous weapons will make the military profession less 'noble' and perhaps even erode the professional identity of military personnel, whose *core* combat skills and responsibilities will be rendered obsolete (Boulanin and Verbruggen 2017: 72). There is another side to this same coin: machines will not be able to go beyond the call of duty and replicate heroic and morally praiseworthy behaviour that humans are capable of. This will lead to the erosion of military values, traditions, and *honour* (Tonkens 2012: 151; Johnson and Axinn 2013: 129 and 135–136).

On the other hand, cultural resistance in the military may be a 'generational phenomenon'. This being the case, it will start to fade when the next generation of personnel, increasingly used to new operational paradigms, takes over (Boulanin and Verbruggen 2017: 72). The culture of heroic sacrifice – be it for one's nation, ideological cause, or comrades – is also exaggerated. In real combat, few combatants will seek fame and glory; most will prefer to complete the mission with as few casualties as possible. What is more, heroism is often associated with heroic *defence*. AWS, ideally suited for the task, will help to avoid unnecessary sacrifice of human life (Noone and Noone 2015: 33).

Cultural *differences* form an important part of the seventh dilemma too. In that respect, it goes beyond what can be thought of as *two* sides of the same coin. The *multiplicity* of meanings and viewpoints is revealed as well, in contrast to all the other dilemmas. One of the most often cited reasons in the calls for the prohibition of autonomous lethal weapons is that they are against human *dignity* and basic human *rights* (HRW 2014). However, there are cultural variations in the ways in which these values can be interpreted. This takes us much further than did the previous dilemma and reveals further weaknesses in the argument itself. Though there is a general consensus that the preservation of human dignity is essential, an Islamic interpretation of what this means will differ in its nuances from that of a Western liberal democracy. For instance, differences concerning the social positions and rights of men and *women* (Sharkey 2019: 81 and 83).

The *universality* of human rights is also jeopardized by variations in the interpretation and implementation of human rights in different parts of the world. Human rights, which would entail restrictions and limitations on certain uses of AI, can undermine a government's economic and political objectives. While human rights were incorporated, or at least mentioned *in passing*, in most governments' national strategies on AI, the *depth* of such provisions varies case by case. For example, Denmark stands out for the specificity of its commitments around human rights and declared them to be part of the *core* of their approach to AI. Germany did note the significance of human rights repeatedly throughout the document. Russia highlighted their great importance too but put specific emphasis on a limited scope of human rights (Bradley and Wingfield 2020: 17 and 21). China acknowledged the importance of personal privacy, intellectual property rights, and moral ethics but dropped the very phrase 'human rights' from the text of its national strategy (Next Generation Plan 2017). There are, in fact, more national strategies that did not foreground human rights but what matters most is that the *language* related to human rights does not guarantee their effective *implementation* (Bradley and Wingfield 2020: 21). Concerns about the militarization of the civilian world, in particular violations of human and civil rights by private security forces, law enforcement, and border control agencies, have become more relevant than ever before (Sharkey 2017: 183). Since the degree to which human rights are protected differs from state to state, a more specific argument has been raised: AWS, soulless and unempathic, can be an ideal tool of dictatorship, repressive violence, oppressive population control, and brutal suppression of peaceful protest (Gubrud 2014: 36; Sharkey 2017: 183). *Slaughterbots*, a fictional video created by advocates of the ban on autonomous lethal weapons, presented a future in which swarms of autonomous drones will be used to assassinate political opponents (Stop Autonomous Weapons 2017).

Concluding remarks

This chapter dissected the controversial nature of AWS. Their characteristics and possible effects were both put under the microscope. Our motivation was to differentiate these *imagined* weapons from other weapons categories in use and under development. We also offered a balanced view of the debate on banning AWS, with their pros and cons presented side by side. Pro-ban arguments have often appealed to common ethical principles and have often been charged with assertive and alarmist language, which makes it harder to get the facts right before pushing for a sweeping ban. Fantasies about evil killer robots and representational idioms that find expression in Hollywood movies have contributed to a general sense of impending danger and misconceptions about technological innovation. By abandoning fictional narratives and one-sided arguments, this chapter helped the reader to form a nuanced and informed opinion. The following chapters will shed light on how killer robots are being *over*-securitized by the Campaign to Stop Killer Robots and the implications thereof. The case is an excellent example of how wide-sweeping overreaction and misrepresentation of technological realities is ineffective, and even detrimental, with regard to achieving the desired outcome.

References

Aleksei Ramm (2021) 'У России Есть Своя Линейка Беспилотников-Камикадзе' ['Russia Has Its Own Line of Kamikaze UAVs']. *Izvestia*, 19 February 2021. Available at: https://iz.ru/1126653/aleksei-ramm/u-rossii-est-svoia-lineika-bespilotnikov-kamikadze.

Altmann J, Sauer F (2017) 'Autonomous Weapon Systems and Strategic Stability'. *Survival* 59(5): 117–142.

Arkin R (2010) 'The Case for Ethical Autonomy in Unmanned Systems'. *Journal of Military Ethics* 9(4): 332–341.

Arkin R (2017) 'A Roboticist's Perspective on Lethal Autonomous Weapon Systems'. In: *Perspectives on Lethal Autonomous Weapon Systems*, UNODA Occasional Papers, No. 30. New York: United Nations Publication.

Arkin R (2018) 'Lethal Autonomous Systems and the Plight of the Non-combatant'. Chapter 15. In: Kiggins R (ed) *The Political Economy of Robots*: 317–326. Switzerland: Palgrave Macmillan – International Political Economy Series.

Army Technology (n.d.) 'T-14 Armata Main Battle Tank'. Accessed 13 September 2021. Available at: https://www.army-technology.com/projects/t-14-armata-main-battle-tank/.

Asaro P (2008) 'How Just Could a Robot War Be?' Part II. In: Briggle A, Waelbers K, Brey PAE (eds) *Current Issues in Computing and Philosophy*. Amsterdam: IOS Press.

Asaro P (2012) 'On Banning Autonomous Weapon Systems: Human Rights, Automation, and the Dehumanization of Lethal Decision-Making'. *International Review of the Red Cross* 94(886): 687–709.

Ayoub K, Payne K (2016) 'Strategy in the Age of Artificial Intelligence'. *Journal of Strategic Studies* 39(5/6): 793–819.

Birnbacher D (2016) 'Are Autonomous Weapons Systems a Threat to Human Dignity?' Chapter 5. In: Nehal Bhuta et al. (eds) *Autonomous Weapons Systems: Law, Ethics, Policy*: 105–121. Cambridge: Cambridge University Press.

Boulanin V, Verbruggen M (2017) *Mapping the Development of Autonomy in Weapon Systems*. Stockholm International Peace Research Institute (SIPRI), November 2017. Available at: https://www.sipri.org/sites/default/files/2017-11/siprireport_mapping_the_development_of_autonomy_in_weapon_systems_1117_1.pdf.

Bradley C, Wingfield (2020) *National Artificial Intelligence Strategies and Human Rights: A Review*. Global Partners Digital, Global Digital Policy Incubator & Stanford Cyber Policy Center, April 2020. Available at: https://www.gp-digital.org/wp-content/uploads/2020/04/National-Artifical-Intelligence-Strategies-and-Human-Rights%E2%80%94A-Review_.pdf.

Brenneke M (2018) 'Lethal Autonomous Weapon Systems and Their Compatibility with International Humanitarian Law: A Primer of the Debate'. Chapter 3. In: Gill T, Geiß R, Krieger H, Paulussen C (eds) *Yearbook of International Humanitarian Law*, Vol. 21: 59–98. Berlin: Springer.

Canning JS (2002) *A Definitive Work on Factors Impacting the Arming of Unmanned Vehicles*. Doc. No. NSWCDD/TR-05/36, Dahlgren Division, Naval Surface Warfare Center.

Carpenter C (2016) 'Rethinking the Political/-Science-/Fiction Nexus: Global Policy Making and the Campaign to Stop Killer Robots'. *American Political Science Association* 14(1): 53–69.

Chatham House (2018) 'Banning Autonomous Weapons Is Not the Answer'. Expert Comment, 2 May 2018. Available at: https://www.chathamhouse.org/2018/05/banning-autonomous-weapons-not-answer.

Chen S (2018) 'China's Plan to Use Artificial Intelligence to Boost the Thinking Skills of Nuclear Submarine Commanders'. *South China Morning Post*. Published 4 February 2018. Updated 6 February 2018. Available at: https://www.scmp.com/news/china/society/article/2131127/chinas-plan-use-artificial-intelligence-boost-thinking-skills.

Davison N (2017) 'A Legal Perspective: Autonomous Weapon Systems under International Humanitarian Law'. In: *Perspectives on Lethal Autonomous Weapon Systems*, UNODA Occasional Papers, No. 30. New York: United Nations Publication.

Defense Post (2020) 'General Atomics to Develop Smart Sensor Technology for MQ9 Reaper Drone'. 27 November 2020. Available at: https://www.thedefensepost.com/2020/11/27/sensor-technology-mq9-reaper-drone/.

Dehuri S, Cho SB, Ghosh S (2011) 'Swarm Intelligence and Neural Networks'. Chapter 1. In: Cho SB et al. (eds) *Integration of Swarm Intelligence and Artificial Neural Network*: 1–21. Singapore: World Scientific.

Dialani P (2019) 'AI Missiles Will Be Developed by the US Army to Locate Their Own Targets'. *Analytics Insight*, 21 August 2019. Available at: https://www.analyticsinsight.net/ai-missiles-will-be-developed-by-the-us-army-to-locate-their-own-targets/.

DoD Directive (2012/2017) *Directive 3000.09: Autonomy in Weapon Systems*. US Department of Defense. Adopted 21 November 2012. Modified 8 May 2017. Available at: https://irp.fas.org/doddir/dod/d3000_09.pdf.

Finn A, Scheding S (2010) *Developments and Challenges for Autonomous Unmanned Vehicles: A Compendium*. Berlin: Springer.

FLI [Future of Life Institute] (2015) *Autonomous Weapons: An Open Letter from AI and Robotics Researchers*. Announced 28 July 2015. Accessed 28 September 2021. Available at: https://futureoflife.org/open-letter-autonomous-weapons/.

Freedberg SJ (2019) 'ATLAS: Killer Robot? No. Virtual Crewman? Yes.' *Breaking Defense*, 4 March 2019. Available at: https://breakingdefense.com/2019/03/atlas-killer-robot-no-virtual-crewman-yes/.

Garcia D (2015) 'Killer Robots: Why the US should Lead the Ban'. *Global Policy* 6(1): 57–63.

Garcia D (2018) 'Lethal Artificial Intelligence and Change: The Future of International Peace and Security'. *International Studies Review* 20(2): 334–341.

Ghosh P (2019) 'Call to Ban Killer Robots in Wars'. *BBC News*, 15 February 2019. Available at: https://www.bbc.com/news/science-environment-47259889.

Gubrud M (2014) 'Stopping Killer Robots'. *Bulletin of the Atomic Scientists* 70(1): 32–42.

Hambling D (2021) 'U.S. Navy Destroys Target with Drone Swarm and Sends a Message to China'. *Forbes*, 30 April 2021. Available at: https://www.forbes.com/sites/davidhambling/2021/04/30/us-navy-destroys-target-with-drone-swarm---and-sends-a-message-to-china/?sh=6a83b0292df1.

Heyns C (2013) *Report of the Special Rapporteur on Extrajudicial, Summary or Arbitrary Executions*. Doc. No. A/HRC/23/47. Available at: https://www.ohchr.org/Documents/HRBodies/HRCouncil/RegularSession/Session23/A-HRC-23-47_en.pdf.

Heyns C (2016) 'Autonomous Weapons Systems: Living a Dignified Life and Dying a Dignified Death'. Chapter 1. In: Nehal Bhuta et al. (eds) *Autonomous Weapons Systems: Law, Ethics, Policy*: 3–19. Cambridge: Cambridge University Press.

Horowitz MC (2016) 'Why Words Matter: The Real World Consequences of Defining Autonomous Weapons Systems'. *Temple International and Comparative Law Journal* 30(1): 85–98.

HRW [Human Rights Watch] (2012) 'Losing Humanity: The Case against Killer Robots'. 19 November 2012. Available at: https://www.hrw.org/report/2012/11/19/losing-humanity/case-against-killer-robots#.

HRW [Human Rights Watch] (2014) 'Shaking the Foundations: The Human Rights Implications of Killer Robots'. 12 May 2014. Available at: https://www.hrw.org/report/2014/05/12/shaking-foundations/human-rights-implications-killer-robots.

HRW [Human Rights Watch] (2015) 'Precedent for Preemption: The Ban on Blinding Lasers as a Model for a Killer Robots Prohibition'. 8 November 2015. Available at: https://www.hrw.org/news/2015/11/08/precedent-preemption-ban-blinding-lasers-model-killer-robots-prohibition.

HRW [Human Rights Watch] (2016) 'Making the Case: The Dangers of Killer Robots and the Need for a Preemptive Ban'. 9 December 2016. Available at: https://www.hrw.org/report/2016/12/09/making-case/dangers-killer-robots-and-need-preemptive-ban.

HRW [Human Rights Watch] (2018) 'Heed the Call: A Moral and Legal Imperative to Ban Killer Robots'. 21 August 2018. Available at: https://www.hrw.org/report/2018/08/21/heed-call/moral-and-legal-imperative-ban-killer-robots.

ICRC [International Committee of the Red Cross] (2014) 'Autonomous Weapons Systems: Technical, Military, Legal, and Humanitarian Aspects'. 1 November 2014. Available at: https://www.icrc.org/en/document/report-icrc-meeting-autonomous-weapon-systems-26-28-march-2014.

Jha UC (2016) *Killer Robots: Lethal Autonomous Weapon Systems: Legal, Ethical and Moral Challenges*. India: Vij Books India.

Johnson AM, Axinn S (2013) 'The Morality of Autonomous Robots'. *Journal of Military Ethics* 12(2): 129–141.

Kalashnikov Media (2018) 'Демонстрация Работы Системы Автоматического Управление Огнем' ['Demonstration of the Automatic Fire Control System']. 1 October 2018. Available at: https://kalashnikov.media/video/technology/demonstratsiya-raboty-sistemy-avtomaticheskogo-upravlenie-ognem.

Kania EB (2018) 'Chinese Sub Commanders May Get AI Help for Decision-Making'. *Defense One*, 12 February 2018. Available at: https://www.defenseone.com/ideas/2018/02/chinese-sub-commanders-may-get-ai-help-decision-making/145906/.

Kastan B (2013) 'Autonomous Weapons Systems: A Coming Legal "Singularity"?' *University of Illinois Journal of Law, Technology and Policy* 1: 45–82.

Keller J (2017) 'The Company that Built the AK-47 Is Developing a Real Life "Terminator"'. *The National Interest*, 18 July 2017. Available at: https://nationalinterest.org/blog/the-buzz/the-company-built-the-ak-47-developiong-real-life-terminator-21578.

Klincewicz (2015) 'Autonomous Weapons Systems, the Frame Problem and Computer Security.' *Journal of Military Ethics* 14(2): 162–176.

Korać ST (2018) 'Depersonalisation of Killing: Towards A 21st Century Use of Force "Beyond Good and Evil"?' *Philosophy and Society* 29(1): 49–64.

Krishnan A (2009) *Killer Robots: Legality and Ethicality of Autonomous Weapons*. Farnham: Ashgate Publishing.

Kruglov A, Ramm A, Dmitriev E (2018) 'Средства ПВО Объединят Искусственным Интеллектом' ['Air Defense Systems Will Be Integrated with the Help of Artificial Intelligence']. *Izvestia*, 2 May 2018. Available at: https://iz.ru/733333/aleksandr-kruglov-aleksei-ramm-evgenii-dmitriev/sredstva-pvo-obediniat-iskusstvennym-intellektom.

Lele A (2017) 'A Military Perspective on Lethal Autonomous Weapon Systems'. In: *Perspectives on Lethal Autonomous Weapon Systems*, UNODA Occasional Papers, No. 30. New York: United Nations Publication.

Leveringhaus A (2016) *Ethics and Autonomous Weapons*. London: Palgrave Macmillan.

Lin P, Bekey G, Abney K (2008) *Autonomous Military Robotics: Risk, Ethics, and Design*. Version: 1.0.8. Prepared for US Department of Navy, Office of Naval Research. California Polytechnic State University. Available at: https://digitalcommons.calpoly.edu/cgi/viewcontent.cgi?article=1001&context=phil_fac.

Liu HY (2012) 'Categorization and Legality of Autonomous and Remote Weapons Systems'. *International Review of the Red Cross* 94(886): 627–652.

Lucas GR (2010) 'Postmodern War'. *Journal of Military Ethics* 9(4): 289–298.

Lucas GR (2014) 'Automated Warfare'. *Stanford Law and Policy Review* 25: 317–339.

Lye H (2020) 'US Army Developing Self-Targeting AI Artillery'. *Army Technology*. Originally published 16 August 2019. Updated 1 June 2020. Available at: https://www.army-technology.com/news/us-army-developing-self-targeting-ai-artillery/.

Matthias A (2004) 'The Responsibility Gap in Ascribing Responsibility for the Actions of Automata'. *Ethics and Information Technology* 6(3): 175–183.

McFarland T (2015) 'Factors Shaping the Legal Implications of Increasingly Autonomous Military Systems'. *International Review of the Red Cross* 97(900): 1313–1339.

Next Generation Plan [New Generation Artificial Intelligence Development Plan] (2017) *Full Translation: China's 'New Generation Artificial Intelligence Development Plan'*. Trans. by Webster G, Creemers R, Triolo P, Kania E. Available at: https://www.newamerica.org/cybersecurity-initiative/digichina/blog/full-translation-chinas-new-generation-artificial-intelligence-development-plan-2017/.

Noone GP, Noone DC (2015) 'Debate Over Autonomous Weapons Systems'. *Case Western Reserve Journal of International Law* 47(1): 25–35.

NYT [The New York Times] (2019) 'A.I. Is Making It Easier to Kill (You). Here's How'. *Youtube*, 13 December 2019. Available at: https://www.youtube.com/watch?v=GFD_Cgr2zho.

O'Connell ME (2014) '21st Century Arms Control Challenges: Drones, Cyber Weapons, Killer Robots, and WMDs'. *Washington University Global Studies Law Review* 13(3): 515–533.

Osborn K (2017) 'Air Force Chief Scientist Confirms F-35 Will Include Artificial Intelligence'. *Defense Systems*, 20 January 2017. Available at: https://defensesystems.com/articles/2017/01/20/f35.aspx.

PAX [PAX for Peace] (n.d.) 'Killer Robots'. Accessed 28 September 2021. Available at: https://paxforpeace.nl/what-we-do/programmes/killer-robots.

Ria Fan [Federal News Agency] (2018) 'Пусть Теперь США Догоняют: Эксперт Рассказал о Прорыве "Калашникова"' ['Let the US Catch Up Now: Expert Told about "Kalashnikov" Breakthrough']. 1 October 2018. Available at: https://riafan.ru/1105275-pust-teper-ssha-dogonyayut-ekspert-rasskazal-o-proryve-kalashnikova.

Ria Novosti (2021) 'Источник: "Армата" Впервые в Истории Сама Нашла Цели без Участия Экипажа' ['Source: For the First Time in History "Armata" Itself Identified Targets Without the Participation of the Crew']. 25 February 2021. Available at: https://ria.ru/20210225/armata-1598859233.html.

Robillard M (2018) 'No Such Thing as Killer Robots'. *Journal of Applied Philosophy* 35(4): 705–717.

Roff H (2015) 'Autonomous or "Semi" Autonomous Weapons? A Distinction Without Difference'. *The Huffington Post*. Published 16 January 2015. Updated 18 March 2015. Available at: https://www.huffpost.com/entry/autonomous-or-semi-autono_b_6487268.

Rosenberg Z (2017) 'Military Drones Can Now Deal with Threats on Their Own'. *Air & Space*, 27 April 2017. Available at: https://www.airspacemag.com/daily-planet/no-pilot-no-problem-180963050/.

Rosert E, Sauer F (2021) 'How (Not) to Stop the Killer Robots: A Comparative Analysis of Humanitarian Disarmament Campaign Strategies'. *Contemporary Security Policy* 42(1): 4–29.

Scharre P (2011) 'Why Unmanned'. *Joint Force Quarterly* 61(Q2): 89–93.

Scharre P (2017) 'A Security Perspective: Security Concerns and Possible Arms Control Approaches'. In: *Perspectives on Lethal Autonomous Weapon Systems*, UNODA Occasional Papers, No. 30. New York: United Nations Publication.

Scharre P (2018) *Army of None: Autonomous Weapons and the Future of War*. New York: W.W. Norton.

Schmitt MN (2013) 'Autonomous Weapon Systems and International Humanitarian Law: A Reply to the Critics'. *Harvard National Security Journal Feature*: 1–37. Available at: https://harvardnsj.org/wp-content/uploads/sites/13/2013/02/Schmitt-Autonomous-Weapon-Systems-and-IHL-Final.pdf.

Sharkey A (2019) 'Autonomous Weapons Systems, Killer Robots and Human Dignity'. *Ethics and Information Technology* 21: 75–87.

Sharkey N (2008a) 'Cassandra or False Prophet of Doom: AI Robots and War'. *IEEE Intelligent Systems* 23(4): 14–17.

Sharkey N (2008b) 'Grounds for Discrimination: Autonomous Robot Weapons'. *RUSI Defence Systems* 11(2): 86–89.

Sharkey N (2010) 'Saying "No!" to Lethal Autonomous Targeting'. *Journal of Military Ethics* 9(4): 369–383.

Sharkey N (2012) 'The Evitability of Autonomous Robot Warfare'. *International Review of the Red Cross* 94(886): 787–799.

Sharkey N (2017) 'Why Robots Should Not Be Delegated with the Decision to Kill'. *Connection Science* 29(2): 177–186.

Sivakumaran S (2012) *The Law of Non-International Armed Conflict*. Oxford: Oxford University Press.

Sparrow R (2007) 'Killer Robots'. *Journal of Applied Philosophy* 24(1): 62–77.

Sparrow R (2009a) 'Predators or Plowshares? Arms Control of Robotic Weapons'. *IEEE Technology and Society Magazine* 28(1): 25–29.

Sparrow R (2009b) 'Building a Better WarBot: Ethical Issues in the Design of Unmanned Systems for Military Applications'. *Science and Engineering Ethics* 15(2): 169–187.

Stanavov A (2020) 'Железная Гвардия: Самые Опасные Боевые Роботы России' ['Iron Guard: The Most Dangerous Fighting Robots in Russia']. *Ria Novosti*. Published 15 October 2017. Updated 3 March 2020. Available at: https://ria.ru/20171015/1506649786.html.

Stop Autonomous Weapons (2017) 'Slaughterbots'. *Youtube*, 13 November 2017. Available at: https://www.youtube.com/watch?v=9CO6M2HsoIA.

Stop Autonomous Weapons (n.d.) 'Development Status'. Accessed 25 September 2021. Available at: https://autonomousweapons.org/development-status/.

Tamburrini G (2016) 'On Banning Autonomous Weapons Systems: From Deontological to Wide Consequentialist Reasons'. Chapter 6. In: Nehal Bhuta et al. (eds) *Autonomous Weapons Systems: Law, Ethics, Policy*: 122–141. Cambridge: Cambridge University Press.

TASS [Russian News Agency] (2018a) 'Ростех: Россия Уже Имеет Наработки по Самообучающемуся Оружию с Искусственным Интеллектом' ['Rostec: Russia Already Has Experience in Self-Learning Weapons with Artificial Intelligence']. 6 June 2018. Available at: https://tass.ru/armiya-i-opk/5268765.

TASS [Russian News Agency] (2018b) '"Революционное Оружие России": Чем "Ураны" Удивили Американские СМИ' ['"Russian Revolutionary Weapons": How "Uran" Series Surprised American Media']. 15 May 2018. Available at: https://tass.ru/armiya-i-opk/5199268.

TASS [Russian News Agency] (2020) 'Russia Begins Trials on Voice Control of Military Robots'. 29 June 2020. Available at: https://tass.com/defense/1172663.

Tonkens R (2012) 'The Case against Robotic Warfare: A Response to Arkin'. *Journal of Military Ethics* 11(2): 149–168.

UNODA [United Nations Office for Disarmament Affairs] (2017) *Perspectives on Lethal Autonomous Weapon Systems*, UNODA Occasional Papers, No. 30. New York: United Nations Publication. Available at: https://www.un.org/disarmament/wp-content/uploads/2017/11/op30.pdf.

Valagin A (2017) 'Боевой Робот "Нерехта-2" Будет Управляться Жестами' ['Fighting Robot "Nerekhta-2" Will Be Gesture Controlled']. *Rossiyskaya Gazeta*, 1 May 2017. Available at: https://rg.ru/2017/05/01/reg-cfo/boevoj-robot-nerehta-2-budet-upravliatsia-zhestami.html.

Vallor S (2016) *Technology and the Virtues: A Philosophical Guide to a Future Worth Wanting*. New York: Oxford University Press.

Wagner M (2014) 'The Dehumanization of International Humanitarian Law: Legal, Ethical, and Political Implications of Autonomous Weapon Systems'. *Vanderbilt Journal of Transnational Law* 47(5): 1371–1424.

Wakefield J (2018) 'South Korean University Boycotted Over "Killer Robots"'. *BBC News*, 5 April 2018. Available at: https://www.bbc.com/news/technology-43653648.

Walsh JI (2015) 'Political Accountability and Autonomous Weapons'. *Research and Politics* 2(4): 1–6.

Zawieska K (2017) 'An Ethical Perspective on Autonomous Weapon Systems'. In: *Perspectives on Lethal Autonomous Weapon Systems*, UNODA Occasional Papers, No. 30. New York: United Nations Publication.

5 Over-securitizing Autonomous Weapons Systems

The Campaign to Stop Killer Robots

Introduction

The aim of this chapter is to shed light on the process of securitization concerning autonomous weapons systems (AWS). The focus is on a coordinated response by civil society to the potential development and use of such weapons – the Campaign to Stop Killer Robots. The difference between 'killer robots', the emotive and resonant term associated directly with the Campaign, and the more general and neutral term 'AWS' is discussed in Chapter 4. Though we disagree with their interchangeability, here we often refer to AWS by the Campaign's generally preferred term 'killer robots', which is best fitted to convey its collective message. The Campaign is modelled on other, previously successful humanitarian disarmament campaigns and has, since its formation in 2012, actively lobbied for 'a pre-emptive and comprehensive ban on the development, production, and use of fully autonomous weapons' (CSKR 2013). It has united almost 200 non-governmental organizations (NGOs) based in various countries, gained broad public support, and successfully dragged into its agenda a good number of state governments, thousands of experts, over 20 Nobel Peace Prize laureates, as well as elements of the United Nations (UN) and the European Union (EU) (CSKR n.d.-a). However, at least for the moment, their collaborative effort to establish a desired legal norm has not succeeded. There is still no international regime formally banning, or even purposefully regulating, AWS.

Our intention is to highlight in a succinct and disciplined manner some of the problems the Campaign and its supporters have encountered in their efforts to securitize and subsequently ban AWS. We foreground the apparently paradoxical effects of what we call *over*-securitization, an area that remains seriously underexplored. In doing so, we rely on – and further refine – the existing securitization literature. In particular, we conceptualize and study how broad-based securitization and multi-layered stigmatization can in some situations, paradoxically, lead away from the desired goal. Two concepts are introduced to define two different yet interrelated mechanisms through which over-securitization operates: *hybridization* and *grafting*. These provide the conceptual basis to analyse the rhetoric and visuals used by varying relevant actors to securitize or, eventually, over-securitize, fully autonomous weapons. Our theoretical contribution is, therefore, to forge an

DOI: 10.4324/9781003045489-8

original way of thinking about the process and existing tools of securitization. Our contribution to the empirical literature stems from the application of securitization, as a method of understanding the logic of social and political construction of threats, to the case of AWS. We demonstrate how the increasingly vociferous, transnationally stretched political campaign has actively created the spectre of danger that is yet flawed in many ways. The findings substantially advance our present understanding of the Campaign's surprising lack of wider political success which, as we show, reflects a tension between *efficiency* and *effectiveness*.

The paradox of over-securitization: Is more necessarily better?

The concept of *securitization* and the original logic of the process were introduced by the Copenhagen School, mainly Barry Buzan and Ole Wæver. Their intention was to broaden the understanding of security from military to non-military threats and introduce the concept of societal security beyond the realm of national security (Buzan, Wæver, and de Wilde 1998: 21–23). Two constitutive elements of their theoretical approach are particularly important for this chapter: the notion of a *securitizing move* and that of a *securitizing actor*. Securitization theory has been further developing ever since. The general tendency central to the present discussion is the redefinition and broadening of our understanding of the *means* of securitization and the *identity* of securitizing actors.

A securitizing move is, according to the original definition by the Copenhagen School, a specific rhetorical structure or discourse – or, more precisely, a *speech act* – that frames an issue as an existential threat, i.e., a security issue (Buzan, Wæver, and de Wilde 1998: 25–26). The central goal of this chapter is the identification and analysis of such speech acts because speech acts have indeed been the main means by which the Campaign has increasingly sought to securitize AWS. Since other means have also been employed, we will draw intellectual resources from other approaches to securitization theory too. There is still more, however, that we can borrow from the Copenhagen School. Wæver (1995) pointed to the performative dimension of the move: 'the utterance itself is *the act* [emphasis added]'. He even defined the process as '[t]he performance of the security act'. Huysmans (2006: 147) accurately spotted the linguistic practice turning from mere description to performative acts as one of the organizing principles in social and political relations. One important caveat is that *performativity* is not the same thing as a theatre performance, but is about the ways in which the discourse is reiterated and networks are created (Ringmar 2019). In this chapter, we systematically illustrate how a particular communication within a distributed network produces and reinforces a shared understanding of insecurity. What deserves special note is that the emergent security issue is not necessarily linked to a real existential threat but has to be constructed and presented as such a threat (Buzan, Wæver, and de Wilde 1998: 24). There have been insights showing that it does not stop with a politically and socially constructed threat. Discourses of insecurity are often about representations of 'danger' with material consequences (Campbell 1992). This concept is important for us too because the kind of threat we examine here has been constructed through, inter alia, discourses and representations of *danger*.

A person or a group that performs the speech act is a securitizing actor. Success depends to a great extent on the *authority position* held by the actor. Only a few actors and groups do in fact have 'the power to define security'. Among them are political leaders, governments, bureaucracies, lobbyists, and pressure groups (Buzan, Wæver, and de Wilde 1998: 31 and 40). Therefore, the Copenhagen School gave us a hint as to the nature of securitizing actors. Wæver (1995: 57) insisted specifically, however, that 'security is articulated only from a specific place, in an institutional voice, [typically] by elites'. Buzan, Wæver, and de Wilde (1998: 21) reaffirmed the principle in their joint book: 'Traditionally, by saying "security," *a state representative* [emphasis added] declares an emergency condition.' The limitation of the approach of the Copenhagen School is that it restricts the actors who can legitimately construct security primarily to political leaders. Also, the Copenhagen School provided initial instruction on how to assess the role the audience plays in constructing insecurity, yet this is another limitation. Success, in their view, depends on the audience being convinced that the issue is an existential security threat. The issue is *securitized* only if and when emergency measures that go beyond standard political procedures are accepted as justified (Buzan, Wæver, and de Wilde 1998: 23–25). The audience's role is, therefore, limited to the mere fact of their accepting the truth of the discourse. Other approaches to securitization theory will allow us to compensate for these limits.

Another major contributor to securitization theory has been the Paris School, situated within the emerging field of international political sociology and mainly represented by Didier Bigo and Thierry Balzacq. Their ideas further advanced the understanding of securitizing actors endowed with the authority to securitize. New actors appeared on the stage who were professionals with expertise in a given field. The Paris School put emphasis on expert security knowledge and the 'authority of statistics' (Bigo 2006). Security professionals and security agencies, the ones who routinely collect and analyse data, were recognized as having the authority to determine what exactly constitutes security (Bigo 2000: 176). Particular attention was drawn to bureaucracies that serve as an 'intermediary' with the central government and are directly involved in the provision of security services (e.g., military and police services, border guards and customs agents, intelligence services, risk assessment experts, etc.) (Bigo 2006). Increasingly, security professionals, both researchers and military personnel, are becoming facilitators for the success of securitization, as we show in this chapter. Bigo (2002: 83) specifically highlighted that, even if NGOs intervene, 'they can do so only by turning professional'. Our analysis here illustrates, among other things, the trend towards professionalization in NGOs, themselves collecting, distributing, and efficiently utilizing professional knowledge. The success of the securitizing move is, consequently, linked to the structural position of the speaker within a *relevant* institution (Bigo 2002: 73–74). Bigo (2006) also noted that such institutions can overstep national boundaries and form transnational bureaucratic links. Transnationalization represents, in his view, yet another source of knowledge and 'symbolic' power. The transnational advocacy Campaign to Stop Killer Robots, which is the focus of this chapter, is perhaps the best illustration of how it can play out in everyday practice. Balzacq (2005: 178–179) listed the very same attributes of authorized security speakers: knowledge, trust, and the speaker's power position, formal or symbolic. One of his key points was that 'effective securitization is power-laden'. In addition, he carved out a more central role for the audience, focusing on 'the *power* [emphasis added] that both speaker and listener bring to the interaction' (Balzacq 2005: 171–172). This insight is still not enough to accurately define the role of the audience for the purposes of this chapter, but it brings us closer to the task.

The conceptualization of the securitizing move itself extends beyond the single speech act, as claimed by the Paris School. Bigo (2002: 65–66) stressed the importance of bureaucratic practices performed by security professionals and involved in the creation of administrative knowledge (e.g., population profiling, risk assessment, statistical calculation, category creation, proactive preparation, etc.). Therefore, practical work, discipline, and expertise are certainly no less important than the discourse (Bigo 2000: 194). Here we concentrate less on bureaucratic practices and more on different discursive frames in play. However, we still note, for example, that the Campaign has also drawn quite heavily upon surveys done by the market research company Ipsos. The most significant observation that comes out of it, however, is different: the Paris School broadened the definition of the securitizing move beyond the single, typically political, speech act. Balzacq (2005: 191) also focused on the 'manner' in which the securitizing actor makes the case for the point and drew attention to two basic principles ensuring ultimate success: 'emotional intensity' and 'logical rigor'. He also reminded us of the role of analogies, metaphors, stereotypes, metonymies, lies, gestures, and even silence as effective tools of persuasion (Balzacq 2005: 172 and 179). In this chapter, we illustrate the significance of such tools for creating the terrifying image of 'killer robots'. The increasingly blurred lines between fact and fiction, rationality and emotion are of particular significance to our understanding of that image. Last, but not least, the Paris School pointed to the existence of numerous stakeholders with heterogeneous profiles who do not share the same logics and are often in competition with each other (Bigo 2006). In such 'power struggles', securitizing actors align to eliminate alternative voices and swing the audience's support towards their preferred course of action (Balzacq 2005: 173 and 179). It is a relevant observation, but a careful analysis of such struggles between the ban Campaign and its opponents is the subject of the next chapter.

Other, especially more recent, supplements to securitization theory contributed to the development of a more sophisticated set of assumptions regarding the authority of securitizing actors and the practice of threat construction. Stritzel (2012: 553) summarized the general tendency as the development away from static understandings of the authority to securitize and single speech acts to more complex processes of authorization and more dynamic representations of existential threat. Pointing out such complexities is one of the key objectives we set for ourselves in this chapter. Vuori (2008: 76) assumed that even those actors who fall outside official authority, but have sufficient 'social capital', can become securitizing actors. Social capital is a special kind of asset available, for example, to NGOs. In this chapter, we deal directly and specifically with the ways in which such capital is being mobilized. Berling (2011: 386) convincingly argued, in addition, that science co-determines the status of the securitizing actor. She called it the 'authority of objectivity' (Berling 2011: 393). Brauch (2009: 94) also drew attention to the significance of scientific 'reputation'. Unsurprisingly, then, there is a strong focus on raising a scientific profile within – and in association with – the Campaign to Stop Killer Robots. This is achieved, inter alia, by joining forces with senior scientists and larger scientific communities. The so-called 'epistemic communities' can indeed draw on their knowledge and expertise to influence the outcome of securitization moves 'through reason and facts on the basis of (objective) knowledge' (Floyd 2020: 12). An epistemic community is a network of professionals from a variety of disciplines with recognized scientific expertise and competence in a particular domain, according to the original conceptualization of the term (Haas 1992: 3). Haas (1992: 18) conceptualized epistemic communities as a separate category, inherently distinct from interest groups, social movements, and bureaucratic coalitions. Even though he equipped us with the appropriate concept, the definition is problematic. It is increasingly difficult to draw definitive dividing lines between scientists, political activists, and activist bureaucrats, as we clearly show here. A closer look reveals two serious problems: scientific facts are selectively used, twisted, or suppressed by political activists to serve their political purpose; and scientists themselves become willingly and directly involved in policy making and political manipulation. Berling (2011: 392) assumed that 'scientific capital' co-determines 'the *hierarchy* [emphasis added] in the field of security and the chances of winning'. Yet the opposite is the case in this chapter. Hierarchies are being replaced with more horizontal heterarchies, as we will demonstrate. Science contributes significantly to the chances of success and successful claims for legitimacy, but scientists do not necessarily play a central role in the process of securitization.

There has, at the same time, been greater awareness that the audience can reinforce the authority of the securitizing actor. Salter (2008: 321–322) conceptualized interactions between the securitizing actor and the audience as 'iterative'. He studied the process of 'audience-speaker *co-constitution* [emphasis added] of authority and knowledge'. McInnes and Rushton (2011: 117) even introduced the idea that original audiences can, at some points themselves, act as securitizing actors. These insights help us conceptualize and, more importantly, analyse a much more active engagement on the part of the audience and the increasingly blurred line between securitizing actors and their target audiences. We call it *hybridization*, as explained below, and show how it works in practice. We also clearly demonstrate that such hybridization has no clear boundaries. Floyd (2020) drew our attention to the role of 'functional actors', often confused with the audience but, in fact, seeking to positively or negatively influence the trajectory of securitization. The concept itself, though originally underdeveloped, is borrowed from the Copenhagen School. Floyd (2020: 10) stressed that media outlets can prioritize certain issues over others, decide how information is relayed, and, therefore, control what becomes public knowledge. Vultee (2011: 77–93) showed practically how the media 'speak security'. Our intention through this text is to once again demonstrate that the media play a crucial role in shaping public threat perception and disseminating stereotypes. The very idea of 'killer robots' caught public attention and imagination through broad media coverage. Foreign politicians can, in turn, either provide or withhold external legitimation, according to Floyd (2020: 10). The most important consequence, as we show, is *peer pressure*, i.e., attempts by the truly pro-ban states to push the blocking coalition towards adopting the ban on AWS. Members employed in the relevant industry can equally facilitate or impede securitization by presenting reasoned arguments for one side or the other (Floyd 2020: 11). We have a perfect illustration here: many technology companies have supported the Campaign to Stop Killer Robots and publicly committed not to engage in the development of AWS. Clearly, such statements have significantly strengthened the spirit of the Campaign. Floyd's (2020: 12) interpretation also suggests that even ordinary people can affect the dynamics of securitization, for example, through social media activism. Here we illustrate how ordinary people can facilitate securitization by partaking in surveys and voicing their concerns about the perceived threat. We also highlight that the Campaign has closely aligned with a broad wave of public concern, which has further reinforced its legitimacy in the eyes of the general public.

Numerous studies have also sought to develop a more nuanced understanding of the core of securitizing moves. First of all, there has been a general awareness that scientific models, data, and facts 'can be mobilized strategically' (Berling 2011: 393). Stritzel (2012: 560) explored the link between power politics and popular culture as 'a principal background of meaning'. In particular, he inquired into the use of pop culture and cultural myths by securitizing actors. Williams (2003: 526–527) suggested that images and other visual representations are also part of 'a broader performative act' and do play a significant role in the process of securitization. This chapter shows how scientific facts and cultural imaginaries, visual and discursive frames are all mobilized simultaneously to *construct* the threat of so-called killer robots. Some scholars have even sought to highlight the supposed links between different security agendas such as the 'migration-terrorism nexus' (Ihlamur-Öner 2019), the 'terrorism-asylum nexus', or even the 'terrorism-immigration-asylum nexus' (Tsoukala 2006: 612 and 618). This is an important, yet undeveloped, argument. It is a good starting point for us to properly conceptualize processes of *grafting* involved in a securitizing move, as further elaborated below. Salter (2012: 934), at the same time, reminds us of the fact that 'securitization is a constant process of struggle and contestation'. In accord with his interpretation, the so-called securitizing *move* in fact consists of 'overlapping … language security games performed by varying relevant actors' (Salter 2012: 931). The same author noted elsewhere (2008: 321) that there are also distinct types of audiences and that the move may be successful with some of them, but not others. It is a reminder for us: systematic hybridization between different types of actors acting at different stages in the securitization process, as we illustrate, does not necessarily lead to greater homogenization. We show it here too.

Building on this rich and diverse body of knowledge, the insights we borrow primarily include the significance of scientific reputation, the varied and dynamic identities of securitizing actors, the increasingly blurred line between securitizing actors and their target audiences, and the diversity of ways in which the threat can be constructed and presented. Yet we also go beyond to inquire into the process of *over*-securitization. There have already been efforts to understand and theorize over-securitization, so the term itself is not new. However, the existing literature approaches the problem from the perspective of *referent objects of security*. Hammerstad (2008: 1–2) was the first to point out that a security issue can 'become over-securitised to the point where it is in danger of creating threats [to the referent object] where before there were none'. Ihlamur-Öner (2019: 210) concurred: 'The securitization of irregular and forced migration has reached to the point that it can be described as over-securitization, which creates more threats where there were none, while putting the lives of migrants and refugee protection at risk.'

We take a different approach still and explore the dynamics of over-securitization from the perspective of *securitizing actors* and *securitizing moves*. In doing so, we distinguish two mechanisms through which over-securitization operates at these two levels: *hybridization* and *grafting*, respectively. Through the former, we demonstrate how complex hybridization of securitizing actors and target audiences produces circulatory, transepistemic, and post-truth configurations of security. When making this argument, we are inspired by Aradau and Huysmans. They accurately determined that 'transepistemic relations create greater symmetry between various knowledges and dilute the superior authority of science in truth telling and factual knowledge about the world' (Aradau and Huysmans 2018: 49). The same authors (2018: 54) defined the condition of post-truth as 'a less hierarchical and more horizontal transversal practice of knowledge creation and circulation'. We have a practical illustration of their statements and argue that, when the tipping point is reached, the diffusion of authority decreases the likelihood of successful securitization. The concept of 'grafting [a new norm onto existing norms]' stands for a well-established legal practice and is borrowed from Price (1998: 617). Here it helps us conceptualize how a new security agenda can be *grafted*, through both discursive and visual representations, onto other security issues of immediate importance and even science fiction imagery. Such grafting techniques, as we show, reinforce the sense of insecurity but, quite counter-intuitively, can in some cases impede instead of facilitate the securitization process. The imprecise focus and reductionist paths make it more difficult to name the threat clearly, which is a precondition for successful securitization. Salter (2012: 938–940) gave an example of what it means if 'the threat remains *vague* [emphasis added]'.

Here lies the paradox: success in broadening the stakeholder base and generating the sense of insecurity does not necessarily mean the success of securitization. In a nutshell, greater *efficiency* does not necessarily translate into greater *effectiveness*. Hence, over-securitization is about securitization efforts that are too intense and too ubiquitous, to the point of becoming counter-productive.

Hybridization

The Campaign to Stop Killer Robots has thus far spearheaded much of the securitization effort in the issue area of AWS. The use of emotive terms such as 'killer robots' was selected as an appropriate strategy from the outset as indicated by the very name of the Campaign. The words 'danger' or 'dangerous' have also been in common use. Campaigners have described the emergence of increasingly autonomous weapons as, for instance, a '*dangerous* [emphasis added] development' (HRW 2020a), a 'destabilizing robotic arms race' (CSKR n.d.-b), and 'a serious global threat' (Amnesty International 2015). Further and more specific arguments in favour of the ban are arranged in a systematic way in Chapter 4. Frightening videos have been created as well to alert people about the imminent danger and make sure they feel a true sense of urgency (e.g., Stop Autonomous Weapons 2017; CSKR 2018a; FLI 2019; PAX n.d.). Contributing to the overall growing sense of urgency are statements of this kind: 'China, Israel, Russia, South Korea, the United Kingdom, and the United States are *investing heavily* [emphasis added] in the development of various autonomous weapons systems' (HRW 2020a). Over the last decade, the Campaign has *de facto* grown into a truly global coalition of international, regional, and national NGOs, technology companies, expert communities, governments, administrative officers, and international organizations (IOs). Hence, there are numerous examples of the successful persuasion of audiences and, most importantly, the transformation of the original audience into an even broader coalition of like-minded securitizing actors.

First of all, the line between political activists and experts has become ever more blurred and difficult to define. This directly challenges the argument that epistemic communities differ fundamentally in their general quality of establishing the scientific validity and facts from interest groups and social movements (Haas 1992: 18). In the studied case, academic researchers and technical experts have, in practice and principle, aligned with and directly contributed to the Campaign. AI and robotics researchers wrote an open letter in 2015. This initiative called for 'a ban on offensive autonomous weapons beyond meaningful human control'. The letter portrays autonomous weapons as 'ideal for tasks such as assassinations, destabilizing nations, subduing populations and selectively killing a particular ethnic group'. It has, to date, been signed by 4,502 researchers (FLI 2015). The global health community launched a similar initiative to 'call for an international ban on lethal autonomous weapons'. Their open letter highlighted that 'lethal autonomous weapons can fall into the hands of terrorists and despots, lower the barriers to armed conflict, and become weapons of mass destruction enabling very few to kill very many'. It has so far been signed by 90 health professionals (FLI n.d.-a). In 2017, the leaders of over a hundred AI and robotics companies signed another open letter urging the UN 'to prevent an arms race in these weapons'. They called attention to the fact that lethal autonomous weapons 'can be weapons of terror, weapons that despots and terrorists use against innocent populations, and weapons hacked to behave in undesirable ways'. The letter was signed by Tesla's Elon Musk and DeepMind's Demis Hassabis and

Mustafa Suleyman, among others (FLI 2017a). As of the time of writing, the military personnel's collective letter calling for 'a ban on the development, use and deployment of autonomous weapons' is also being prepared (CSKR n.d.-c). The gap between political activists and experts is closing even faster as NGOs engage in collecting, distributing, and efficiently using professional knowledge. Human Rights Watch, for example, reviewed the precursors to fully autonomous weapons and presented a sound legal analysis to justify the call for preventive action in its 2012 report called 'Losing humanity: the case against killer robots' (HRW 2012). In 2019, PAX released a research report titled 'Slippery slope: the arms industry and increasingly autonomous weapons'. The report provided an overview of recent developments in unmanned technologies and applications of AI (PAX 2019).

The relevant industries have also contributed, and very heavily, to the Campaign. The initiative was supported by many companies and individuals working in the field of robotics and AI: since 2018, 247 organizations and 3,253 individuals pledged to 'neither participate in nor support the development, manufacture, trade, or use of lethal autonomous weapons'. In their pledge, lethal autonomous weapons are described, inter alia, as 'powerful instruments of violence and oppression'. The pledge calls upon governments and government leaders 'to create a future with strong international norms, regulations and laws against lethal autonomous weapons'. Among the signatories are Google DeepMind, Clearpath Robotics, Silicon Valley Robotics, GoodAI, and TeslaVision Corporation (FLI n.d.-b).

There is also an emerging like-minded bureaucratic coalition spreading at both the regional and global levels. In 2013, the Special Rapporteur of the UN Human Rights Council (UNHRC), Christof Heyns, called on 'all states to declare and implement national moratoria on at least the testing, production, assembly, transfer, acquisition, deployment and use of LARs [lethal autonomous robots]'. In his view, allowing such robots to kill people 'may denigrate the value of life itself' and 'seriously undermine the ability of the international legal system to preserve a minimum world order' (Heyns 2013: 20–21). The UN Secretary-General (UNSG), António Guterres, delivered the following message at the Web Summit in 2018: 'machines that have the power and the discretion to take human lives are politically unacceptable, are morally repugnant, and should be banned by international law' (Guterres 2018). In its resolution of 12 September 2018, the European Parliament (EP) urged the member states 'to work towards the start of international negotiations on a legally binding instrument prohibiting lethal autonomous weapon systems'. The same document cautioned, among other things, that 'lethal autonomous weapon systems have the potential to fundamentally change warfare by prompting an unprecedented and uncontrolled arms race' (EP 2018).

Peer pressure has also proved extremely powerful as an advocacy tool and has been growing constantly. Dozens of countries have since called for an outright ban on fully autonomous weapons: Pakistan (2013), Ecuador (2014), Egypt (2014), the Holy See (2014), Cuba (2014), Ghana (2015), Bolivia (2015), the State of Palestine (2015), Zimbabwe (2015), Algeria (2016), Costa Rica (2016), Mexico (2016), Chile (2016), Nicaragua (2016), Panama (2016), Peru (2016), Argentina (2016), Venezuela (2016), Guatemala (2016), Brazil (2017), Iraq (2017), Uganda (2017),

Austria (2018), Djibouti (2018), Colombia (2018), El Salvador (2018), Morocco (2018), Jordan (2019), and Namibia (2019). China's ban call (2018) is limited to the use of fully autonomous weapons only and does not cover their development and production (CSKR 2020a). These countries have themselves taken active part in the securitization of the threat of AWS. For example, in November 2015, Iraq associated fully autonomous weapons with 'an arms race which could have catastrophic results'. Colombia called such weapons 'a military and legal *threat* [emphasis added]' in December 2016. Brazil warned in November 2017 that certain weapons with autonomous capabilities 'will prove to be incompatible with international humanitarian law and international human rights law'. In October 2018, El Salvador stated that 'a machine that has the responsibility to decide about a person's life … raises great ethical and legal challenges' (cited in HRW 2020a). In April 2018, the Holy See put particular emphasis on the moral side of the problem in the following statement: '[a]n autonomous weapons system could never be a *morally responsible subject* [emphasis added]' (cited in PAX 2018: 16). The 'moral tutelage' by the Holy See is of importance in light of its unique 'moral authority' (Eyffinger 1999: 77–88).

Even those countries whose governments have not necessarily been supportive of the ban have often contributed to the securitization of AWS. For example, in May 2014, the Czech Republic warned that lethal autonomous weapons 'could pose a serious threat for civilians'. In November 2014, Ireland also expressed concern at the 'eventual use of these technologies outside of traditional combat situations, for example in law enforcement'. Kuwait stated in October 2015 that such weapons systems 'pose moral, humanitarian, and legal challenges'. In October 2017, Myanmar explicitly characterized lethal autonomous weapons as 'a *security issue* [emphasis added]'. Finland concluded in November 2017 that the 'development of weapons and means of warfare where humans are completely out of the loop would pose serious risks from the ethical and legal viewpoint' (cited in HRW 2020a).

Now we turn our attention to the special status and role of the International Committee of the Red Cross (ICRC). The ICRC has international legal personhood, i.e., a 'legal status vis-à-vis states in international law'. It is, at the same time, a recognized 'humanitarian expert' (Mathur 2011: 182, 2017: 19). First and foremost, however, it is the guardian of international humanitarian law (IHL), or the so-called Geneva Conventions (Forsythe 2005: 13; Mathur 2011: 182, 2017: 4 and 95). Since this is the case, the ICRC has long been at the 'intersection' of disarmament and IHL (Mathur 2017: 15). Normative and diplomatic facilitation by the ICRC has been observed consistently, for instance, in the case of AWS. It has, since 2015, urged states 'to establish internationally agreed limits on autonomous weapon systems to ensure civilian protection, compliance with international humanitarian law, and ethical acceptability' (ICRC 2021a). Peter Maurer, the President of the ICRC, stressed in his recent speech that 'the use of autonomous weapons to target human beings should be ruled out'. He explained his motive in saying so as follows: 'The potential humanitarian consequences are concerning for the ICRC. These weapon systems raise serious challenges for compliance with international humanitarian law' (cited in ICRC 2021b).

The media, traditionally perceived as the medium between securitizing actors and their audiences, have generously and actively contributed to the threat rhetoric and the killer robots discourse. Such titles abound: 'Terminator or Robocop?' (*The Economist*, May 2013); 'The Rise of the Killer Robots – And Why We Need to Stop Them' (*CNN*, October 2015); 'Killer Robots: New Reasons to Worry About Ethics' (*Forbes*, January 2016); 'Is "Killer Robot" Warfare Closer Than We Think?' (*BBC*, August 2017); 'Killer Robots Must Be Banned But "Window to Act is Closing Fast," AI Expert Warns' (*The Independent*, November 2017); 'Killer Robots Are Coming: Scientists Warn UN Needs Treaty to Maintain Human Control Over All Weapons' (*The National Post*, November 2017); 'Stop the Rise of the "Killer Robots," Warn Human Rights Advocates' (*The Washington Post*, November 2017); 'Killer Robots Will Only Exist If We Are Stupid Enough to Let Them' (*The Guardian*, June 2018); 'Why We Need a Pre-emptive Ban on "Killer Robots"' (*The Huffington Post*, August 2018); 'Killer Robots Aren't Regulated. Yet.' (*The New York Times*, December 2019); '"Killer Robots" and AI Could Wipe Out Humanity, Report Warns' (*The Telegraph*, August 2020). Broad media coverage has indeed been identified by the Campaign as one of the contributing mechanisms (CSKR 2018b: 26–42).

It is also interesting to observe how ordinary people, the audience in its most traditional sense, become actively engaged in the process of securitization as well. They participate in polls and surveys which are then used by pro-ban advocates to publicly reinforce their position and the view that their policies are justified. Human Rights Watch announced in 2019, based on the survey conducted in December 2018 by Ipsos, that 61 per cent of respondents from 26 countries are opposed to the development of killer robots (HRW 2019). Based on the new survey conducted by Ipsos in December 2020, the Campaign publicly declared in 2021 that opposition to killer robots remains strong and that more than three in five people responding to a new online survey in 28 countries oppose the use of fully autonomous weapons (CSKR 2021). Human Rights Watch revealed, rather surprisingly, in the report titled 'Children Vote to Stop Killer Robots' that interest in the Campaign is 'growing across the world, *especially among children* [emphasis added]' (HRW 2020b). The organization clearly sent the message to further open the gateway to a new trend of recognizing children as political contributors, a trend set by Swedish climate activist Greta Thunberg.

The previous paragraphs clearly demonstrate that different actors did participate at different points in the creation of baseline knowledge about the threat of AWS. The relationship between a large number of experts in different fields, bureaucrats, policy advocates, mass media, and the general public have become more symmetrical. The securitizing 'actor' here is, therefore, a *hybrid* construction. There is a repeated circulation of the same or similar arguments, iterated and reiterated continually. Truth has become more relational with such a diverse set of actors involved in defining security. Objective facts are, therefore, often reduced to a range of subjective interpretations. Science is increasingly hijacked by political activists of whom many are scientists themselves, which is detrimental to conducting unbiased and impartial scientific research. It means circulatory, transepistemic, and post-truth knowledge underlies the understanding of what constitutes a security threat in the case of AWS.

In fact, one may argue that the case is not so different from many other securitization processes in the field of arms control and disarmament. We can indeed cite numerous examples of when a wide range of actors were involved directly or indirectly in the securitization process (Hynek and Solovyeva 2020). However, the case we analyse here is almost unique (except perhaps cyber weapons, cf. Stevens 2019: 284). While it is relatively easy to define nuclear, biochemical, and laser weapons, for instance, the term 'killer robots' is notoriously difficult to define. So, the diffusion of authority to different actors reduces the likelihood of success simply because it is getting more difficult to name the threat clearly and put across one single scientific proven message to the target audience.

Having systematically illustrated overlapping security games, we identify the three most important challenges that arise from them: no clear definition of the so-called killer robots; no certainty as to whether (or to what extent) they already exist; and no certainty as to whether or not AI is central to their primary function. For example, the following definition is given by Human Rights Watch: 'Fully autonomous weapons, also known as "killer robots," would be able to select and engage *targets* [emphasis added] without meaningful human control' (HRW n.d.). However, this definition is broad enough to cover, for example, autonomous air and missile defence systems which have long been accepted as legitimate. Amnesty International provided a more specific definition: 'Killer robots are weapons systems which, once activated, can select, attack, kill and injure *human targets* [emphasis added] without a person in control' (Amnesty International 2015). PAX reports that '[k]iller robots do not yet exist' (PAX n.d.). In a section on killer robots on their website dedicated to humanitarian disarmament, Harvard Law School's Armed Conflict and Civilian Protection Initiative added a sense of urgency to the definition. In their interpretation, killer robots are 'currently under development, … moving rapidly closer to reality' (ACCPI n.d.). Another source of direct contribution to the Campaign contains somewhat contradictory information: 'In reality, weapons which can autonomously select, target, and kill humans *are already here* [emphasis added]' (Stop Autonomous Weapons n.d.-a). The BBC summarized one of the key initiatives in this issue area as follows: 'A group of scientists has called for a ban on the development of weapons *controlled* [emphasis added] by artificial intelligence (AI)' (Ghosh 2019). PAX defines 'killer robots … as weapons which, once activated, using sensors *and/or* [emphasis added] artificial intelligence, will be able to operate without meaningful human control over the critical functions' (PAX 2017/2018: 6). These are only selected examples but they do indicate incoherence, ambivalence, and uncertainty. Yet it is important to define the threat as precisely as possible to close interpretive loopholes. We acknowledge, however, that the main goal has been a fairly clear issue from the very outset: 'Life and death decisions should not be delegated to a machine' (CSKR n.d.-d). Nevertheless, differences in nuance and emphasis may be detrimental to the Campaign's ability to succeed when such a complex weapons category is concerned.

Grafting

There have been further problems at the level of threat construction. We interpret and problematize them here under the rubric of *grafting*. The concept, as already explained above, was introduced by Price. To be still more precise, he defined it as 'the mix of genealogical heritage and conscious manipulation involved in … normative rooting and branching'. What is of particular importance, according to his theorization, is how well a new norm 'resonates' with already established norms. For example, he argued that the effort to delegitimize anti-personnel landmines was *grafted* onto a viable chemical weapons taboo, the laws of war, and IHL (Price 1998: 628–629). We demonstrated elsewhere that the same logic and processes did, to a greater or lesser extent, underlie other cases of arms control and disarmament as well (Hynek and Solovyeva 2020). However, in all these cases, a new norm is typically grafted onto international legal treaties, i.e., legal consensuses that already exist. Chapter 6 discusses, inter alia, how the same methods are being pursued by the Campaign to Stop Killer Robots. But, to reinforce the sense of political urgency, the Campaign took it one step further: their security agenda is also being *grafted*, both discursively and visually, onto other security issues of immediate importance and science fiction imagery. Below we propose a series of arguments that highlight problematic aspects of such multilayered stigmatization.

One of the key narrative tropes is that of biased AI. What has been foregrounded are gender and race biases, as well as the lack of legal regulations over the use of personal data. For instance, the Campaign warned that human biases will end up influencing intelligent algorithms and publicly stated that 'achieving a ban on fully autonomous weapons or killer robots is a feminist issue' (CSKR 2020b). In this context it was also stressed that lethal autonomous weapons 'increase the risk of targeted violence against classes of individuals, including ethnic cleansing and genocide' (Stop Autonomous Weapons n.d.-b). Another trusted source provides virtually the same information: 'Autonomous weapons are ideal for … selectively killing a particular ethnic group' (FLI 2015). Increasingly one also hears the argument that autonomous weapons 'could become powerful instruments of violence and oppression, especially when linked to surveillance and data systems' (FLI n.d.-b). The Canadian component of the Campaign to Stop Killer Robots provides the following, more general, explanation:

> Killer robots will mostly require artificial intelligence to function, and artificial intelligence requires data to "think." For those of us whose personal data has been available on the internet since we were children, this is a massive risk to our safety and freedom.
>
> (Osler 2021)

But it should not be forgotten that a heightened focus on racism, gender-based discrimination, and personal data safety has been uneven across regions and countries.

However fair and rightful the concerns, placing particular emphasis on these issues can further alienate some parties (e.g., China) from the negotiation process.

Broad, sometimes inaccurate or problematic, generalizations have also been made. For instance, there has been too much focus on the current arms race in AI. The social imaginary of '[a]n *artificially intelligent* [emphasis added] system tasked with autonomous targeting' (Sauer 2016) is certainly part of what makes killer robots fearsome. So arguments of this sort have surfaced: 'Starting a military AI arms race is a bad idea, and should be prevented by a ban on offensive autonomous weapons beyond meaningful human control' (FLI 2015). The Asilomar AI Principles, proposed by AI experts in 2017, are also formulated in a way that identifies the current arms race in AI directly with autonomous killing machines: 'AI Arms Race Principle: An arms race in lethal autonomous weapons should be avoided' (FLI 2017b). The same concerns have even been explicitly presented as involving a clear and present danger. The UNSG, António Guterres, stated, for example, at the Web Summit in 2018: 'The weaponization of artificial intelligence is a *serious danger* [emphasis added]' (Guterres 2018). It is, however, a reductionist trap. Such a narrow definition of the race puts aside other uses of AI military systems and networks, as shown in Chapter 3. Inaccurate representations of current R&D efforts make it easier for the other side, represented mainly by the world leaders in AWS development, to dismiss pro-ban arguments altogether.

The killer robots rhetoric is also grounded in the discourse on weapons of mass destruction (WMD). For example, health professionals highlighted in their aforementioned open letter that 'lethal autonomous weapons can … become weapons of mass destruction enabling very few to kill very many' (FLI n.d.-a). The Future of Life Institute also define lethal autonomous weapons as 'a new class of weapon of mass destruction' (FLI n.d.-c). The arms control advocacy viral video *Slaughterbots* visually demonstrated how killer drone swarms turn into robotic weapons of mass destruction (Stop Autonomous Weapons 2017). It is yet another attempt to take a short cut that is, however, not necessarily the best solution. The analogy between WMD and AWS is deeply problematic but it is rather an attempt to re-create the emotional appeal of WMD. WMD-based language is, in most cases, pervasive enough to stigmatize the weapons in question as immoral and unacceptable. Enemark (2011) concluded, however, and in this we concur, that the term WMD is 'misleading from a technological viewpoint'. He argued that it 'obscures the paramount threat of nuclear weapons, exaggerates the destructive power of chemical weapons, and is unhelpful or counterproductive when used in the context of biological weapons'. One can safely argue that the term is similarly unproductive, and may even be counterproductive if not applied carefully, in the case of AWS. In particular, it bears the risk of over-simplification in that case. Chapter 4 makes clear that there are many proposed benefits to the *actual use* (as opposed to the benefits that *deterrence* yields, for instance, in the case of nuclear weapons) of AWS. This is, at the very least, what distinguishes AWS from all other categories of WMD. Falling into the reductionist trap of such umbrella terms and hastily dismissing substantive counterarguments, pro-ban activists risk jeopardizing the success of the whole enterprise.

Fear of killer robots is also stoked by their extremely stereotyped presentation in science fiction, often invoked to symbolize and visually represent danger. Media coverage of the Campaign and autonomous weapons has broadly featured images of terrifying humanoid military robots, as seen in science fiction films such as *The Terminator* (e.g., Walsh 2015; Devlin 2018). *The Terminator* has indeed become a central metaphor in the killer robots discourse. In one of the BBC reports, virtually the same visual representation was even accompanied by the following words: '"Killer robots" may seem like something from a sci-fi film, but reality is catching up' (Smith 2017). In a *Forbes* report, a very similar image was captioned as '[t]he reality of the rise of autonomous weapons systems' (Pandya 2019). These examples clearly illustrate the importance of visuals in threat construction and presentation. Amnesty International also highlighted the possible link between science fiction and autonomous weapons: '"Killer Robots" will not be a thing of science fiction for long' (Amnesty International 2015). What needs to be stressed, however, is that the Campaign has repeatedly tried to dispel the illusion of killer robots that look like the Terminator. One of their earliest joint statements cited roboticist Noel Sharkey as saying that '[k]iller robots are not self-willed "Terminator"-style robots' (CSKR 2013). AI-enabled weapons, even if still under development, do not really have anything in common with the Terminator, as shown in Chapter 3. However, as also spotted by Carpenter (2016: 60), there is a direct contradiction between what the Campaign claims and how it is represented, fairly widely, by other actors directly or indirectly involved in the cultivation of killer robots as a threat. Misrepresentations of the threat – be they intentional or unintentional – do have an adverse effect on the securitization process because they make it even easier for the other side to lightly dismiss arguments made by the Campaign.

Concluding remarks

This chapter uncovered the paradox related to securitization efforts in a broader sense and the one specific case studied here: greater *efficiency* does not necessarily translate into greater *effectiveness*. On one hand, it allowed us to fill the existing gap in the securitization literature. As we made clear, the problem of over-securitization has heretofore received scant attention. We argued and showed that the continuum of securitization, i.e., the progression from normalcy to the recognition of an existential threat and the adoption of extraordinary measures, is far more complex than that. It is no less important to inquire into securitization efforts that begin as useful and gradually become counter-productive. Here we offered two analytical tools with which to address such problems as soon as they are discovered: the concept of *hybridization* and that of *grafting*. We demonstrated how the increasingly hybridized, more diverse, and seemingly more politically powerful authority of the ones who perform the securitizing act paradoxically leads them away from the desired goal. The chapter also discovered numerous traps related to threat construction and presentation. Grafting a new representation of threat onto a range of other forms of perceived threats is not necessarily helpful, as we show. If discursive and visual frames are not carefully selected, the call for multilateral regulatory action might seem stronger at first glance but will, also quite paradoxically, have a lower chance of materializing.

On the other hand, when identifying the paradox, we grasped and systematically explained the Campaign's surprising lack of wider political success. We comprehensively illustrated that the point has been reached where its unstinted and highly efficient efforts are actually becoming counter-productive. No doubt the Campaign succeeded in not only popularizing the threatening image of killer robots, but also in broadening the base of support for the ban. Simultaneously though, it has failed in its main purpose while falling into reductionist traps, entering several politically contested terrains at the same time, and being enmeshed in fictions and overlapping security games performed by different types of involved actors. What we observe, as a result, is a split between two different tracks: the discourse of disarmament advocates, imbalanced and dominated by a single approach, and the much more complex realities of military R&D, as discussed in detail in Chapter 3. The imbalance in the discourse becomes apparent if one thoroughly familiarizes oneself also with the dilemmas presented in Chapter 4. In the next chapter, we will examine how these two tracks meet, at least indirectly, and the unfortunate, but predictable, consequences of their mutual insulation for arms control.

References

ACCPI [Armed Conflict & Civilian Protection Initiative] (n.d.) 'Killer Robots'. *Humanitarian Disarmament*. Accessed 21 September 2021. Available at: https://humanitariandisarmament.org/issues/killer-robots/.

Amnesty International (2015) 'Ten Reasons Why It's Time to Get Serious about Banning "Killer Robots"'. 12 November 2015. Available at: https://www.amnesty.org/en/latest/news/2015/11/time-to-get-serious-about-banning-killer-robots/.

Aradau C, Huysmans J (2018) 'Assembling Credibility: Knowledge, Method and Critique in Times of "Post-Truth"'. *Security Dialogue* 50(1): 40–58.

Balzacq T (2005) 'The Three Faces of Securitization: Political Agency, Audience and Context'. *European Journal of International Relations* 11(2): 171–201.

Berling TV (2011) 'Science and Securitization: Objectivation, the Authority of the Speaker and Mobilization of Scientific Facts'. *Security Dialogue* 42(4–5): 385–397.

Bigo D (2000): 'When Two Become One: Internal and External Securitisations in Europe'. Chapter 8. In: Kelstrup M, Williams MC (eds) *International Relations Theory and the Politics of European Integration: Power, Security and Community*: 171–204. London: Routledge.

Bigo D (2002): 'Security and Immigration: Toward a Critique of the Governmentality of Unease'. *Alternatives* 27: 63–92.

Bigo D (2006) 'Globalized (In)Security: The Field and the Ban-Opticon'. Chapter 2. In: Bigo D, Tsoukala A (eds) *Terror, Insecurity, Liberty: Illiberal Practices of Liberal Regimes after 9–11*: 10–48. London: Routledge.

Brauch HG (2009) 'Securitizing Global Environmental Change'. Chapter 4. In: Brauch HG et al. (eds) *Facing Global Environmental Change*: 65–102. Berlin: Springer.

Buzan B, Wæver O, de Wilde J (1998) *Security: A New Framework for Analysis*. Boulder: Lynne Rienner.

Campbell D (1992) *Writing Security: United States Foreign Policy and the Politics of Identity*. Minneapolis: University of Minnesota Press.

Carpenter C (2016) 'Rethinking the Political/-Science-/Fiction Nexus: Global Policy Making and the Campaign to Stop Killer Robots'. *American Political Science Association* 14(1): 53–69.

CSKR [Campaign to Stop Killer Robots] (2013) 'Urgent Action Needed to Ban Fully Autonomous Weapons: Non-governmental Organizations Convene to Launch Campaign to Stop Killer Robots'. 23 April 2013. Available at: https://www.stopkillerrobots.org/wp-content/uploads/2013/03/KRC_LaunchStatement_23Apr2013_fnl.pdf.

CSKR [Campaign to Stop Killer Robots] (2018a) 'No Country Would Be Safe from Fully Autonomous Weapons'. *Youtube*, 5 April 2018. Available at: https://www.youtube.com/watch?v=qiJTq11kqdw.

CSKR [Campaign to Stop Killer Robots] (2018b) 'Convention on Conventional Weapons Group of Governmental Experts Meeting on Lethal Autonomous Weapons Systems and Meeting of High Contracting Parties United Nations, Geneva, November 2017'. Report on Activities, 26 February 2018. Available at: https://www.stopkillerrobots.org/wp-content/uploads/2018/02/CCW_Report_Nov2017_posted.pdf.

CSKR [Campaign to Stop Killer Robots] (2020a) 'Country Views on Killer Robots'. 11 March 2020. Available at: https://www.stopkillerrobots.org/wp-content/uploads/2020/03/KRC_CountryViews_11Mar2020.pdf.

CSKR [Campaign to Stop Killer Robots] (2020b) 'Killer Robots, Feminism…And a Feminist Foreign Policy?' 13 February 2020. Available at: https://stopkillerrobots.medium.com/killer-robots-feminism-and-a-feminist-foreign-policy-d55309c60fae.

CSKR [Campaign to Stop Killer Robots] (2021) 'Opposition to Killer Robots Remains Strong – Poll'. 28 January 2021. Available at: https://www.stopkillerrobots.org/news/poll-opposition-to-killer-robots-strong/.

CSKR [Campaign to Stop Killer Robots] (n.d.-a) 'Who Wants to Ban Fully Autonomous Weapons'. Video. Accessed 4 October 2021. Available at: https://www.stopkillerrobots.org.

CSKR [Campaign to Stop Killer Robots] (n.d.-b) 'The Problem'. Accessed 23 September 2021. Available at: https://www.stopkillerrobots.org/learn/#problem.

CSKR [Campaign to Stop Killer Robots] (n.d.-c) 'Military and Killer Robots'. Accessed 1 November 2021. Available at: https://www.stopkillerrobots.org/military-and-killer-robots/.

CSKR [Campaign to Stop Killer Robots] (n.d.-d) 'Less Autonomy. More Humanity'. Accessed 1 November 2021. Available at: https://www.stopkillerrobots.org/.

Devlin H (2018) 'Killer Robots Will Only Exist If We Are Stupid Enough to Let Them'. *The Guardian*, 11 June 2018. Available at: https://www.theguardian.com/technology/2018/jun/11/killer-robots-will-only-exist-if-we-are-stupid-enough-to-let-them.

Enemark C (2011) 'Farewell to WMD: The Language and Science of Mass Destruction'. *Contemporary Security Policy* 32(2): 382–400.

EP [European Parliament] (2018) *European Parliament Resolution on Autonomous Weapon Systems*. Doc. No. 2018/2752(RSP). Available at: https://www.europarl.europa.eu/doceo/document/TA-8-2018-0341_EN.pdf?redirect.

Eyffinger A (1999) *The 1899 Hague Peace Conference: The Parliament of Man, the Federation of the World*. Hague: Kluwer Law International.

FLI [Future of Life Institute] (2015) *Autonomous Weapons: An Open Letter from AI and Robotics Researchers*. Announced 28 July 2015. Accessed 23 September 2021. Available at: https://futureoflife.org/open-letter-autonomous-weapons/.

FLI [Future of Life Institute] (2017a) *An Open Letter to the United Nations Convention on Certain Conventional Weapons*. Accessed 23 September 2021. Available at: https://futureoflife.org/autonomous-weapons-open-letter-2017/.

FLI [Future of Life Institute] (2017b) 'Asilomar AI Principles'. Accessed 4 October 2021. Available at: https://futureoflife.org/ai-principles/.

FLI [Future of Life Institute] (2019) 'Why We Should Ban Lethal Autonomous Weapons'. *Newsletter*, March 2019. Accessed 15 October 2021. Available at: https://futureoflife.org/2019/04/09/fli-march-2019-newsletter/?cn-reloaded=1.

FLI [Future of Life Institute] (n.d.-a) *Lethal Autonomous Weapons: An Open Letter from the Global Health Community*. Accessed 23 September 2021. Available at: https://futureoflife.org/medical-lethal-autonomous-weapons-open-letter/.

FLI [Future of Life Institute] (n.d.-b) *Lethal Autonomous Weapons Pledge*. Accessed 5 October 2021. Available at: https://futureoflife.org/2018/06/05/lethal-autonomous-weapons-pledge/.

FLI [Future of Life Institute] (n.d.-c) 'Lethal Autonomous Weapons Systems'. Accessed 15 October 2021. Available at: https://futureoflife.org/lethal-autonomous-weapons-systems/.

Floyd R (2020) 'Securitisation and the Function of Functional Actors'. *Critical Studies on Security*. Available at: DOI: 10.1080/21624887.2020.1827590.

Forsythe DP (2005) *The Humanitarians: The International Committee of the Red Cross*. Cambridge: Cambridge University Press.

Ghosh P (2019) 'Call to Ban Killer Robots in Wars'. *BBC News*, 15 February 2019. Available at: https://www.bbc.com/news/science-environment-47259889.

Guterres A (2018) 'Remarks at 'Web Summit' by United Nations Secretary-General'. Lisbon, 5 November 2018. Available at: https://www.un.org/sg/en/content/sg/speeches/2018-11-05/remarks-web-summit.

Haas PM (1992) 'Introduction: Epistemic Communities and International Policy Coordination'. *International Organization* 46(1): 1–35.

Hammerstad A (2008) *Securitisation as a Self-fulfilling Prophecy: Refugee Movements and the North-South Security Divide*. Political Science Association Annual Conference, Swansea, April 2008.

Heyns C (2013) *Report of the Special Rapporteur on Extrajudicial, Summary or Arbitrary Executions*. Doc. No. A/HRC/23/47. Available at: https://www.ohchr.org/Documents/HRBodies/HRCouncil/RegularSession/Session23/A-HRC-23-47_en.pdf.

HRW [Human Rights Watch] (2012) 'Losing Humanity: The Case against Killer Robots'. 19 November 2012. Available at: https://www.hrw.org/report/2012/11/19/losing-humanity/case-against-killer-robots.

HRW [Human Rights Watch] (2019) 'Poll Shows Strong Opposition to "Killer Robots"'. 22 January 2019. Available at: https://www.hrw.org/news/2019/01/22/poll-shows-strong-opposition-killer-robots.

HRW [Human Rights Watch] (2020a) 'Stopping Killer Robots: Country Positions on Banning Fully Autonomous Weapons and Retaining Human Control'. 10 August 2020. Available at: https://www.hrw.org/report/2020/08/10/stopping-killer-robots/country-positions-banning-fully-autonomous-weapons-and.

HRW [Human Rights Watch] (2020b) 'Children Vote to Stop Killer Robots'. 9 June 2020. Available at: https://www.hrw.org/news/2020/06/09/children-vote-stop-killer-robots.

HRW [Human Rights Watch] (n.d.) 'Killer Robots'. Accessed 1 October 2021. Available at: https://www.hrw.org/topic/arms/killer-robots.

Huysmans J (2006) *The Politics of Insecurity: Fear, Migration and Asylum in the EU*. New York: Routledge.

Hynek N, Solovyeva A (2020) *The Logic of Humanitarian Arms Control and Disarmament: A Power-Analytical Approach*. London: Rowman and Littlefield.

ICRC [International Committee of the Red Cross] (2021a) 'ICRC Position on Autonomous Weapon Systems'. 12 May 2021. Available at: https://www.icrc.org/en/document/icrc-position-autonomous-weapon-systems.

ICRC [International Committee of the Red Cross] (2021b) 'Peter Maurer: "We Must Decide What Role We Want Human Beings to Play in Life-and-Death Decisions During Armed Conflicts"'. 12 May 2021. Available at: https://www.icrc.org/en/document/peter-maurer-role-autonomous-weapons-armed-conflict.

Ihlamur-Öner SG (2019) 'Delinking the Migration-Terrorism Nexus: Strategies for the De-Securitization of Migration'. *Perceptions* 24(2–3): 195–224.

Mathur R (2011) 'Humanitarian Practices of Arms Control and Disarmament'. *Contemporary Security Policy* 32(1): 176–192.

Mathur R (2017) *Red Cross Interventions in Weapons Control*. Lanham: Lexington Books.

McInnes C, Rushton S (2011) 'HIV/AIDS and Securitization Theory'. *European Journal of International Relations* 19(1): 115–138.

Osler T (2021) 'The Start of Something Good'. Mines Action Canada & Stop Killer Robots Canada. 21 January 2021. Available at: https://stopkillerrobots.ca/2021/01/21/the-start-of-something-good/.

Pandya J (2019) 'The Weaponization of Artificial Intelligence'. *Forbes*, 14 January 2019. Available at: https://www.forbes.com/sites/cognitiveworld/2019/01/14/the-weaponization-of-artificial-intelligence/?sh=3775f7453686.

PAX [PAX for Peace] (2017/2018). 'Where to Draw the Line: Increasing Autonomy in Weapon Systems – Technology and Trends'. Published November 2017. Updated April 2018. Available at: https://paxforpeace.nl/media/download/pax-report-where-to-draw-the-line.pdf.

PAX [PAX for Peace] (2018) 'Crunch Time: European Positions on Lethal Autonomous Weapon Systems'. November 2018. Available at: https://paxforpeace.nl/media/download/pax-rapport-crunch-time.pdf.

PAX [PAX for Peace] (2019) 'Slippery Slope: The Arms Industry and Increasingly Autonomous Weapons'. November 2019. Available at: https://paxforpeace.nl/media/download/pax-report-slippery-slope.pdf.

PAX [PAX for Peace] (n.d.) 'Killer Robots'. Accessed 28 September 2021. Available at: https://paxforpeace.nl/what-we-do/programmes/killer-robots.

Price R (1998) 'Reversing the Gun Sights: Transnational Civil Society Targets Land Mines'. *International Organization* 52(3): 613–644.

Ringmar E (2019) 'The Problem with Performativity: Comments on the Contributions'. *Journal of International Relations and Development* 22(4): 899–908.

Salter MB (2008) 'Securitization and Desecuritization: A Dramaturgical Analysis of the Canadian Air Transport Security Authority'. *Journal of International Relations and Development* 11(4): 321–349.

Salter MB (2012) 'Securitization of US-Canada Border in American Political Discourse'. *Canadian Journal of Political Science* 44(4): 929–951.

Sauer F (2016). 'Stopping "Killer Robots": Why Now Is the Time to Ban Autonomous Weapons Systems'. *Arms Control Association*, October 2016. Available at: https://www.armscontrol.org/act/2016-09/features/stopping-%E2%80%98killer-robots%E2%80%99-why-now-time-ban-autonomous-weapons-systems.

Smith M (2017) 'Is "Killer Robot" Warfare Closer than We Think?' *BBC News*, 25 August 2017. Available at: https://www.bbc.com/news/business-41035201.

Stevens T (2019) 'Global Code: Power and the Weak Regulation of Cyberweapons'. Chapter 13. In: Hynek N, Ditrych O, Stritecky V (eds) *Regulating Global Security: Insights from Conventional and Unconventional Regimes*: 271–295. Cham: Palgrave Macmillan.

Stop Autonomous Weapons (2017) 'Slaughterbots'. *Youtube*, 13 November 2017. Available at: https://www.youtube.com/watch?v=9CO6M2HsoIA.

Stop Autonomous Weapons (n.d.-a) 'Slaughterbots are Here'. Accessed 1 November 2021. Available at: https://autonomousweapons.org/.

Stop Autonomous Weapons (n.d.-b) 'The Risks of Lethal Autonomous Weapons'. Accessed 1 November 2021. Available at: https://autonomousweapons.org/the-risks/.

Stritzel H (2012) 'Securitization, Power, Intertextuality: Discourse Theory and the Translations of Organized Crime'. *Security Dialogue* 43(6): 549–567.

Tsoukala A (2006) 'Democracy in the Light of Security'. *Political Studies* 54(3): 607–627.

Vultee F (2011) 'Securitization as a Media Frame: What Happens When the Media "Speak Security"'. Chapter 4. In: Balzacq T (ed) *Securitization Theory: How Security Problems Emerge and Dissolve*: 77–93. London: Routledge.

Vuori J (2008) 'Illocutionary Logic and Strands of Securitization: Applying the Theory of Securitization to the Study of Non-Democratic Political Orders'. *European Journal of International Relations* 14: 65–99.

Wæver O (1995) 'Securitization and Desecuritization'. Chapter 3. In: Lipschutz RD (ed) *On Security*: 46–86. New York: Columbia University Press.

Walsh T (2015) 'The Rise of the Killer Robots – And Why We Need to Stop Them'. *CNN*, 26 October 2015. Available at: https://edition.cnn.com/2015/10/26/opinions/killer-robots-walsh/index.html.

Williams MC (2003) 'Words, Images, Enemies: Securitization and International Politics'. *International Studies Quarterly* 47(4): 511–532.

6 Operations of Power in Autonomous Weapons Systems Regulation

Introduction

The previous chapter examined processes of the political construction of autonomous weapons systems (AWS) by policy advocacy networks. It made clear that political efficiency did not always go hand in hand with effectiveness, as demonstrated by our focus on processes of over-securitization. This chapter builds on it and extends the analysis further by showing what happens when political advocacy and humanitarian ethics encounter established global politico-economic structures and arms control diplomacy. In particular, it provides the reader with a plastic and comprehensive understanding of the intersection of humanitarian action and great power politics. In contrast to much of the existing literature, we abstract from actor-to-actor interactions only and focus on the operations of power, both direct and diffuse. This allows us to capture and analyse how broader ethical discourses, national security considerations, economic interests, and pre-existing structural asymmetries meet and what the implications are for arms control. AWS, as we show, is yet another emerging technology that animates the existing and sedimented politico-economic arrangements and, simultaneously, the next frontier for political activism. Our goal is to shed light on the prospects of a regulatory, or even prohibitory, regime in the issue area of AWS.

The most hotly debated attribute of AWS is machine autonomy with respect to life-and-death decisions. Given potential humanitarian risks, critics call for a *preventive*, or rather *preemptive*, global ban on their development, production, and use (FLI 2015; HRW 2016). Others conversely argue that such risks might rather be associated with a blanket prohibition on advanced autonomy in weapons systems (Anderson and Waxman 2012: 39–45; Arkin 2018: 321). Individual governments' preferences are divided by a deep rift. Some have either endorsed a full-fledged ban or considered a limited ban at least, while others have been either sceptical to such restrictions or explicitly opposed them (CSKR 2020).

There is a growing body of literature that seeks to study various aspects of security regulation in the age of artificial intelligence (AI) and AWS. Among the issues that have been raised are current trends in government AI regulation vis-à-vis those in international AI regulation (Turner 2019); positive and negative implications of AI for the broader issue area of arms control (Din 1987);

DOI: 10.4324/9781003045489-9

challenges of 'framing' of intelligent weaponry for related arms control practices (Wallach and Allen 2013); assessment of autonomous military vehicles and dangers associated with their use under the 'criteria' of preventive arms control and prohibition (Altmann 2009); challenges that AWS pose to current governance norms (Koppelman 2019); comparative insights into the likelihood of a complete ban on AWS and evaluation of alternative regulatory pathways (Crootof 2015; Rosert and Sauer 2020); and lessons for AI regulation from preceding arms control experiences and paradigms (Maas M. 2018). However, the above literature lacks a comprehensive scrutiny of the dynamics and conditions that have determined the prospect of banning or, at least, effectively regulating AWS. Research on law and ethics has also intensified: both ethical and legal questions come in many shapes and hues in the AWS debate. This body of knowledge includes analyses of how AWS fit within existing international norms, how they challenge their implementation, how they shape normative expectations, and whether the formulation of new norms is necessary (Bode and Huelss 2018). The discussion could be enriched by a systematic disaggregation and explanation of the ethical *and* legal intensity of the emerging regulatory framework.

This chapter seeks to fill these gaps. There are two clearly opposing parties: the vociferous and assertive pro-ban movement and the blocking coalition.[1] The former is primarily represented by the Campaign to Stop Killer Robots, whose establishment and 'ban killer robots' mantra sparked a debate on AWS in political circles. The latter put together the world's leading countries in AWS research and development (R&D) that oppose a ban on AWS, despite still ambiguous policy statements, and can collectively prevent its passing at the Convention on Certain Conventional Weapons (CCW). There are other types of involved actors caught between the two extremes. Since the Campaign primarily put together a coalition of non-governmental organizations (NGOs), there are other actors who, while aligned with its message and goals, do not 'officially' belong to it (Rosert and Sauer 2020: 15–16). Among such explicit, often active, supporters are like-minded governments (e.g., African states), experts (e.g., the International Committee of the Red Cross (ICRC), the Future of Life Institute), and bureaucrats in intergovernmental organizations (IOs) (e.g., the UN Secretary General (UNSG), the UN Human Rights Council (UNHRC), etc.). There are also silent sympathizers as, for example, public surveys sometimes reveal. The situation is similar with the blocking coalition. Though their opposition may be most pronounced, more actors are in fact involved in AWS R&D and focused on exploring the potential benefits of AWS. Closely aligned are academics, scientists, and lawyers actively supplying expertise on strategic importance, military utility, ethical and legal compliance of AWS, as well as private companies working in close collaboration with governments in AWS R&D (e.g., in China). A degree of implicit support is also in place: there are countries (such as Italy) and arms producing companies that continue to invest in AWS R&D but do not clearly position themselves in the debate (Haner and Garcia 2019: 334; PAX 2019; CSKR 2020). The range of different positions adds to the complex socio-political reality that we study, especially as a complete ban is not the only possible course of action. A regulatory treaty of some sort,

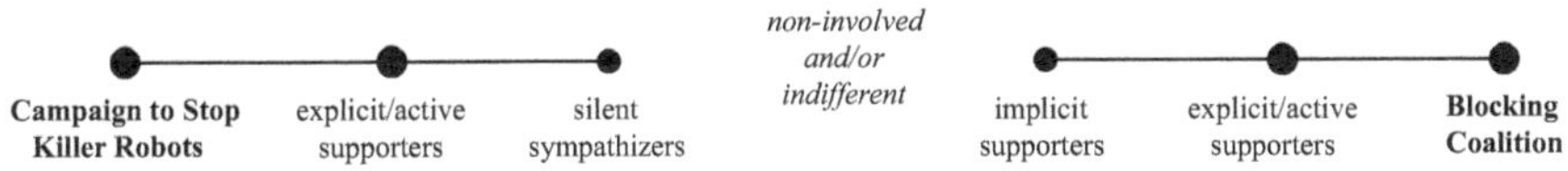

Figure 6.1 Actorness in the debate on banning AWS

domestic laws and policies, nonbinding resolutions, common understandings, professional guidelines, and codes of conduct are possible (Crootof 2015: 1897–1903). The possibility of a political declaration has also been considered (Acheson 2017: 2, 2018: 1). While reflecting on these nuances, we utilize a dichotomy as a heuristic tool as the identified range does not fundamentally challenge the pro- vs anti-ban logic still dominating the AWS debate (Figure 6.1).

Our framework of analysis draws on a *power-analytical approach* to studying international security regimes. It refines it to expound on the nexus between ethics and law and, more generally, on synergies among four conceptual types of power. Here, international security regimes are understood as complex aggregates of normative expectations and legal rules, conditioned by the relationships and interactions among different social actors. Even though there is no global regime formally banning, or even purposefully regulating AWS, this approach is well suited to reflect on attempts at establishing one and the existing resistance. More importantly, it allows the study of multiple discursive, normative, legal, cultural, humanitarian, institutional, political, economic, strategic, military, and technological dynamics as well as various interstices between them as part of a joint investigation. This is because it cuts through the three waves of regime theorization through the conceptualization of power. Ultimately, it enables an understanding of the conditions, interests, and processes that have shaped the prospects of a ban, as well as the ethical *and* legal intensity of the emerging regulatory framework. For a more nuanced assessment, parallels are repeatedly drawn between the selected case and other regulatory and prohibitory precedents, including respective efforts.

Towards security regulation: Legal underbrush and 'severe' politicization

There is currently no global regime formally banning, or even purposefully regulating, fully autonomous (lethal) weapons systems. However, two tightly intertwined strands of developments are relevant for interpreting advancement on this front. First, this chapter sheds light on the legal underbrush concerning AWS security regulation. Second, the success of what might be termed the 'severe politicization' (Emmers 2013: 137) of AWS is put under the microscope. With regard to the former, two aspects are of importance. It is certainly expected that AWS should fit within the readily established legal framework. Primarily, these are International Humanitarian Law (IHL), International Human Rights Law (IHRL), and the Charter of the United Nations (UN) that regulate the use of force, the principle of political responsibility, and the legality of weapons (Liu 2012: 638; Garcia 2015:

60; Brehm 2017). Also falling in this category are the Draft Rules proposed by the ICRC which have proposed the prohibition of weapons with harmful, including uncontrollable and unforeseen, effects on the civilian population (Mathews 2001: 992). The other relevant dimension of the legal backdrop is the emerging framework regulating the use of cyberweapons and its implications for AWS. Such key international mechanisms involve the Wassenaar Arrangement on Export Controls for Conventional Arms and Dual-Use Goods and Technologies as well as the Council of Europe Convention on Cybercrime (Stevens 2019: 271). Both concern, inter alia, digital information, software, computers, computer networks, and telecommunications (CoE 2001; WA 2018). AWS are often discussed with reference to cyberspace, in particular cyber-vulnerability, cyberwarfare, cybercrime, or cyberterrorism (Liu 2012: 648; Schmitt 2013; Klincewicz 2015: 163–170; Noone and Noone 2015: 33; Sharkey 2017: 183). This is because AWS similarly feature (networks of) computers or software-controlled systems operating across the digital and the physical (Dehuri, Cho, and Ghosh 2011: 2; McFarland 2015: 1326–1328; Um 2019: 1–3).

Going beyond the prevailing legal background, the Campaign to Stop Killer Robots has actively lobbied for a complete, issue-oriented ban on AWS. Formed in 2012 as a coalition of NGOs, it has grown into a transnational political advocacy network, or a 'global coalition', as per their self-designation (CSKR n.d.). Cutting through the thicket of legal, ethical, moral, philosophical, political-strategic, scientific, and military discourses, their agenda has deeply penetrated the political scene and the domain of security regulation. Their thrust has spread across global, regional, and local levels. Issue-specific and ad hoc calls, pledges, reports, open letters, directives, resolutions, and declarations have been produced by (groups of) individuals with diverse expertise, NGOs, technology companies, segments of global and regional IOs, and certain governments (e.g., NWI 2014; PAX 2014a; Conn 2018). Since 2013, AWS have had a place on the UN arms control agenda and the UN's CCW framework has served as the principal venue for negotiating a legal instrument on AWS. After three informal expert meetings held to address AWS in 2014, 2015, and 2016, a Group of Governmental Experts (GGE) was established at the CCW. With the GGE meeting regularly since 2017, the AWS debate has gained a 'formal status'. Although the pro-ban movement is dissatisfied with the 'slow pace' of the GGE process, most state parties still see the CCW as the 'appropriate venue' for addressing AWS (Rosert and Sauer 2020: 12 and 17). At this moment, any decision on banning or regulating AWS would be mainly based on GGE recommendations.

The aforementioned legal and political elements represent mere steps towards AWS security regulation but do not furnish one. Given the lessons from the case of cyberweapons, the applicability of more general international laws to regulate a 'weaponized code' may be severely contested (Stevens 2019: 287–288). The legal framework regulating cyberweapons is fragmented and only features piecemeal measures (Stevens 2019: 271). At the same time, despite the great success of giving AWS a political footing, the issue has not become truly 'securitized'. This is because securitizing actors have not yet been successful in convincing states to adopt extraordinary, issue-specific measures beyond standard political procedures (Emmers 2013: 133–135).

A power-analytical approach to international security regimes

In conceptual terms, this chapter builds upon the notion of an *international security regime*. It implies a set of 'principles, rules, and norms that permit nations to be restrained in their behaviour in the belief that others will reciprocate' (Jervis 1982: 357). This conceptual framework allows us to study a regime as an aggregate of both ethical and legal building blocks. The *power-analytical approach* is deployed to dissect the dynamics and conditions that have determined the prospects and limitations of a regulatory, or even prohibitory, regime of this sort in the domain of AWS (Hynek 2018a). In line with this approach, the definition of an international security regime can be further refined as a complex aggregate of normative expectations and legal rules, conditioned by the relationships and interactions among different social actors.

This approach strikes a balance by cutting through the three waves of theorizing regimes within the discipline of IR, namely the neo-neo-convergence regime theory, the cognitivism, and the radical constructivist/post-structuralist understanding (Hynek 2017). The consideration of the multiplicity of power relations is utilized as the means to bring them together and organize regime analysis. Barnett and Duvall's (2005: 48) conceptualization of a 'taxonomy of power', the most sophisticated existing theorization of power, is taken as the basis. It renders four types of power: productive, structural, compulsory, and institutional. First, in relation to expressions of power, one can make a distinction between *interaction* and *constitution*; and second, in connection to the specificity of social relations of power, *direct* and *diffuse* relations can be identified (Barnett and Duvall 2005: 45–48).

As a form of constitutive and diffuse social relations, *productive power* permeates systems of meaning and knowledge. It is used to analyse the discursive and cognitive framing of AWS and related political subjectivities with a regulatory effect. Through the lens of *structural power*, which epitomizes constitutive and direct social relations, power hierarchies and associated interests are examined. This power configuration is utilized to define the key stakeholders in the domain of AWS in terms of the distribution of objectives, obligations, and benefits. It helps to capture structural positions and capacities of the involved actors in direct relation to one another. *Compulsory power* captures a series of interactions through which actors come into direct relations with one other. It allows us to study how the discerned stakeholders in the issue area of AWS are able to control and potentially alter one another's behaviour outside of the working institutional framework. *Institutional power* in turn offers insight into the mediatory power of involved institutions and the nature of diffuse interactions among actors. It allows us to inquire into the availability of venues and decisional rules as well as lines of involvement and responsibility within them and configurations of path-dependence. These are the factors that shape outcomes with respect to institutional limits on AWS and are crucial to consider (Barnett and Duvall 2005: 48–57; Hynek 2018a).

Bourne's (2018: 444) notion of a 'topology of power' is additionally utilized to avoid statism and capture how these forms of power interrelate and cross-pollinate. This allows us to move from 'critical, mutually exclusive, and exhaustive distinctions' (Barnett and Duvall 2005: 43) to 'dynamism that underlies this diversity' (Bourne 2018: 444). We see this analytical move as crucial to grasping the linkage between ethics and law, embedded with complex, synergizing configurations of power. However, this study takes a further step in theorizing the intertwined workings of these four power configurations: it shows how they are not alike in terms of their explanatory value. Wider normative discourses, as well as their historical and structural conditionings, are seen as particularly relevant to regime analysis. To be more specific, this study illustrates how everything starts with productive power, in particular with the workings of *ethical forces*. It is necessary to clarify that forces within the broad ideational structure of ethics are contingent upon and prone to manifold cognitive and affective manipulations. The transposition from the structure of ethics into the legal regulatory realm (IHL, weapons regulation dynamics, etc.) is a matter of successful transfer and consequent *legal encapsulation* of discoursive ethical arguments. One of the key translating and activizing mechanisms between what are otherwise separate structures of ethics and legal regulation in the issue area of weapons regulation is the existing or prospective research, development, and associated funding, all embedded within the 'global arms transfer and production system' (Krause 1992: 205–215). It epitomizes the real or imagined military materialization against which a mix of broad, ethically anchored, and stigmatizing discourses as well as specific legal principles and norms can be utilized. This captures how motions within structural power might kick in as a trigger. In complementing stigmatization production and politics, there is a second-order mechanism that operates solely at the level of legal structures. It is the mechanism of legal 'grafting' (Price 1998: 628) which allows for forging novel legitimizing interlinkages between emerging norms and previously institutionalized norms, themselves products of successful transposition and encapsulation of ethical discursive artefacts. This intellectual advancement accounts for the workings of institutional power in shaping the contours of productive power. Consequently, the constitutive power of ethically anchored discourses works on balance with oppositional narratives, also falling under the confines of productive power, to either maintain or reverse structural power. This interaction heavily relies on mechanisms of direct influence rendered by compulsory power. Such a complex power interplay may result in the production of normative 'hooks' and, at best, the modification of ethical artefacts into a set of more cogent and specific *legal principles and norms*. The latter come into being within the confines of institutional power, hence we argue that institutional power has, to a large degree, a synthetic quality despite its distinctive logics and processes. The proposed theorization allows us to trace and disaggregate the ethical *and* legal intensity of AWS regulation, to provide a robust explanation and to reflect on the prospects. All of this is graphically nuanced to order empirical analysis (Figure 6.2).

Productive power: Nascent stigmatization of a 'politicized' category

This section focuses on the discursive and cognitive framing of AWS with a regulatory effect. Particular attention is paid to the role of *ethics*. The goal is to assess the constructive power and prominence of stigmatizing narratives on balance with competing narratives. Light is shed on the constitutive role of technologies and differences in their interpretation, legal conditions, production and transfer peculiarities, economic disparities, intellectual traditions and habitual practices, ethical and cultural standards.

The principal objective of pro-ban actors has been to portray AWS as illegitimate military instruments. Generally speaking, they build the case upon the convergence of deontology and consequentialism. From the perspective of the former, they denounce the automation of lethal decision-making as unacceptable by definition in moral and ethical terms (Asaro 2012: 708–709; Zawieska 2017: 49). This deontological viewpoint is reinforced by multi-dimensional consequentialist reasoning. Here we give some examples, while a more detailed analysis of arguments in favour of the ban is presented in Chapter 4. AWS are condemned by critics as 'operat[ing] in a lawless zone' (Kastan 2013: 47). In the first place, they reputedly threaten to breach IHL and IHRL, including the principle of individual and state political responsibility and accountability for violations (Asaro 2008, 2012: 692–693). Similar and closely related concerns are raised with respect to the UN Charter norm of (non-)use of force, the principle of legal review of (new) weapons codified in the additional protocol to the Geneva Conventions, as well as the Martens Clause introduced in the second Hague Convention and the Geneva Conventions Additional Protocols (Gubrud 2014: 40). The premise of non-compliance in the case of AWS particularly lies with what is termed the 'dehumanization of killing' (Wagner 2014: 1410) and the 'depersonalization of war' (Klincewicz 2015: 163). The former implies depriving combat of healthy human emotions and judgement (Asaro 2012: 701; Johnson and Axinn 2013: 136; Wagner 2014: 1415; Sharkey 2017: 180). The latter stands for both the depersonalization of responsibility (Heyns 2016: 12) and the objectification of enemy (non-)combatants (Sharkey 2012: 788; Korać 2018: 54). All of this is condemned as a threat to the fundamental values of human dignity and human life (Sparrow 2007: 68; Heyns 2013: 6, 2016: 3; Johnson and Axinn 2013: 134; Garcia 2015: 58–61). The violation of such values by lethal autonomous weapons has even been identified as an infringement on the 'norm of civilization' (Rosert and Sauer 2019: 373; for the original and more general argument, cf. Price 1995).

Interestingly, the debate over the morality of drone warfare, especially in light of recent technological advances in facial recognition, has been framed in opposite terms as targeted killings, i.e., the (hyper-) personalization of war (Dunlap 2014: 109–111; Schulzke 2017: 56–57). The inability of AWS to conform to legal rules and normative standards is often attributed to the unpredictable character of such technologies. It derives from the impossibility of coding full-fledged intelligence in the first place, inherent software imperfections, and vulnerability to cyber attacks (Lin, Bekey, and Abney 2008: 20; Asaro 2012: 691; Klincewicz 2015: 167–170; Noone and Noone 2015: 33). What exacerbates the case is a warning that autonomous weapons might evolve into a mass product or 'the Kalashnikovs of tomorrow'. This technological turn is described with reference to a 'revolution in warfare' (FLI 2015), albeit the alarm is sounded that this one is different from earlier military revolutions (Heyns 2013: 5) and even represents a radical departure from strategic affairs (Payne 2018: 218). The most radical fears feature robots 'destroying humankind' (Lin, Bekey, and Abney 2008: 63).

To reinforce all these narratives, pro-ban actors have, to a large extent, anthropomorphized the debate on autonomous weapons. This is how *killer robots* have emerged as 'a historical imaginary of the 21st century' (Karppi, Böhlen, and Granata 2016: 120). The very notion of 'killer robots' relies on a horrifying representation of AWS as a machine version of human soldiers (e.g., Devlin 2018). Chapter 3 made clear that such a narrow, one-sided representation of artificially intelligent weapons neglects other uses of AI in military systems and networks, especially those hidden from public view. Nevertheless, the Campaign has made the decision to put emphasis on the unique capacity of autonomous killing machines, even if they are non-existent, as shown in Chapter 4. It is because the narratives of barbaric killer robots have strong visceral appeal and resonate well with the large audience of non-experts and civilians. The underlying rationale is to convey 'a simple and dramatic message' (Rosert and Sauer 2020: 18; for the general notion of short causal chains, see Keck and Sikkink 1998). It is, however, not necessarily the best strategy: further problems associated with the killer robots discourse are thoroughly discussed in Chapter 5.

Given all the risks, those working to ban killer robots (CSKR n.d.) have drawn a sharp dividing line between themselves and ban-resisting actors. Having portrayed the removal of meaningful human control over the use of lethal force as unethical, irresponsible, immoral, and illegal (Asaro 2012: 695–709), they have particularly labelled this a moral red line (CSKR 2019) or a moral threshold (HRW 2016). Discussing the 'precedent' of a preemptive ban, i.e., the one on blinding laser weapons (BLW), Human Rights Watch (2018) has recalled the notions of 'civilized nations' or 'civilization' and, on balance, that of 'barbarity'. Resembling the fate of nuclear and chemical weapons, this discursively (re)produced rift again delineates the ingroup of 'normals' or the 'civilized' (Price 1995: 95; Sauer and Reveraert 2018: 438). With AWS portrayed also as an attribute of rich, elaborate economies, the stigmatization of the outgroup is further reinforced via fears of imbalanced and asymmetric warfare (Asaro 2008: 62–63; Garcia 2018: 339).

Structural conditions of emergence have paved the way for the deep embeddedness of this discourse. The general condition of possibility is the traditional distinction between human and machine with respect to control over life and death (Karppi, Böhlen, and Granata 2016: 111–112). The deeply rooted history of technological advancement has produced a strong imaginary of weapons as 'passive tools' (Bourne 2012: 142). It consolidates the dominant logic that 'guns don't kill people, people do' (Latour 1999: 174) and problematizes the hypothetically reversed logic that 'guns don't kill people, cyborgs do' (Bourne 2012: 141).

The prohibitionary character of killer robots is also co-constituted by deeply rooted science fiction and popular culture images and stereotypes (Carpenter 2016). Countless sci-fi films and novels play a remarkable role in creating the image of a robot that could potentially become self-aware and disobey its programming (Krishnan 2009: 7–8). This imaginary mainly reflects Western cultural representations of automata depicted in movies such as *The Terminator* (1984) or *I, Robot* (2004) (Karppi, Böhlen, and Granata 2016: 111). What facilitates the stigmatization of AWS along these lines is that laws are often imperfect, incomplete, and require human interpretation and judgement to apply to real world situations (Asaro 2012: 700–705). The principle of human dignity and the right to life serve particularly well to divert attention from the military utility of AWS (Garcia 2015: 60–61; Heyns 2016: 8–10). These norms have become ever more prominent as a result of a macro-shift in the systemic ethical force after the end of the Cold War. This transformation entails the replacement of *sovereignty* by *human rights* as the central force of compassion. The concept of security is consequently fixed upon a human dimension previously within the sovereign purview of states (Krause 2002: 260). Echoing the nineteenth-century humanitarian dispositif, there has occurred the (re)constitution of an individual as a referent object of security (Hynek 2018b).

A few precedents played a crucial role in embedding post-Cold War ethical forces and humanitarian dispositifs into weapons regulation dynamics. Bans previously imposed upon certain military instruments such as biological and chemical weapons, anti-personnel landmines (APL), and cluster munitions (CM) encouraged uncompromising ambitions in the area of humanitarian disarmament (Hynek 2018b). Though AWS rely on more complex, dual-use, and strategically advantageous (combinations of) technologies, the CCW Protocol IV *preemptively* banning BLW within the CCW set a particularly important precedent (HRW and IHRC 2015; HRW 2018). The limited advancement of legal control of cyberweapons, as detailed above, in turn provides a starting point and precursor principles for AWS security regulation. The advocacy underwriting the case of cyberweapons even embeds and complements the same polemic on AWS. Since existing international laws are barely applicable to the (insecure) nature of code and the Internet, issue-oriented legal principles are seen as indispensable in addressing such transnational challenges (Stevens 2019: 272, 280, and 287). The key role of precedents is in norm-setting: roughly the same set of legal, especially humanitarian, principles served as the basis for the Chemical Weapons Convention, the Mine Ban Treaty, the Convention on Cluster Munitions, etc. (Hynek 2018b). Though yet failing to score a global security regime, similar concerns are raised with respect to regulating cyberweapons (Stevens 2019). As we showed, the stigmatization of AWS is to a large extent *stimulated* by this repeated practice of employing similar foundations of legal arguments and humanitarian language.

Alongside the growing appeal of legal grafting, it is countries' acceleration and increasing investments in AWS R&D that *triggered* the pro-ban movement. Early indications of this trend were clearly evident at the beginning of the 2010s (Haner and Garcia 2019). At around the same time that it became increasingly conceived that AWS could be 'developed within 20 to 30 years', the Campaign to Stop Killer Robots appeared (HRW 2012). These developments came to be seen as especially worrisome against the background of a general rise of the economic within the political. Such a structural transformation revealed a tension between broader changes in ethics and other structural forces, namely conflict and economy (Hynek 2018b). The *stratified* global arms transfer system, involving domestic sales from companies to governments, (illicit) trade, and co-production arrangements, grabbed particular attention (Krause 1992: 205–215; Stavrianakis 2010: Chap. 2). In view of the dual-use nature of AWS-enabling technologies (FLI 2015; Altmann and Sauer 2017: 132), the context of civilian technologies becoming increasingly useful to the military adds to the overall picture (Smith and Udis 2003: 102–103). Such contextual conditions have also been thought of in connection with declining production costs, including recent technological advancements in 3D printing (Haner and Garcia 2019: 331). Also, they have been associated with the global trend of turning code into a weapon (Stevens 2019: 289). Similar concerns have already been raised in relation to drones supposedly creating 'unethical relationships of civil-military technology sharing' (Schulzke 2017: 55).

Well thought-through, coherent, and deeply embedded, the prohibition discourse has gained a strong foothold. It is fair to conclude that 'a stigma is already becoming attached to the prospect of removing meaningful human control from weapons systems and the use of force' (CSKR 2019). However, the central underlying goal of a global prohibition regime for (lethal) autonomous weapons has not been achieved. Nor has any form of issue-oriented legal regulation been adopted. Which explanations might be discerned at the stage of scrutinizing the discourse? Though grounded in materiality and often likened to the cases of biological, chemical, gunpowder, and nuclear weapons (Singer 2009: 179 and 203; Heyns 2013: 5; FLI 2015), killer robots are less a physical entity than a 'politicized' concept (Karppi, Böhlen, and Granata 2016: 111). Except for precursor systems of various complexity, fully autonomous (lethal) weapons do not exist thus far since they have not yet been extensively developed and have never been deployed (Noone and Noone 2015: 28; Walsh 2015: 2; Sauer 2016; Horowitz 2016b: 7). The very notion of killer robots renders the issue 'futuristic' (Rosert and Sauer 2020: 13 and 18). It means that advocacy has overrun practice. This in no small part accounts for the failure to properly define the object of stigmatization at this moment (Horowitz 2016a). In general, the nature of autonomy in AWS derives from an AI-enabled *process* that can take any form, perform cognitive functions, and carry out complex tasks without human intervention. There are concerns that true autonomous weapons might even challenge the distinction between weapons, methods of warfare, and warriors (Liu 2012: 628–629 and 636–637; Heyns 2013: 6). The situation also echoes and exacerbates a few setbacks on the way to cyberweapons governance. The dual-use, digital, and unconventional (physical) nature of technologies may make their global prohibition or over-regulation neither desirable nor possible in terms of verification, compliance, and enforcement (Stevens 2019: 272, 280, and 288). Stigma politics apparently lacks a meaningful practical basis with respect to AWS, even though R&D of increasingly autonomous weapons is well underway, as shown in Chapter 3. Besides that, the prohibition discourse seriously downplays the advantages of AWS. This triggers a thrust of firmly grounded scepticism, sometimes even resistance. Many have drawn attention to the military utility and strategic importance of AWS, as well as illusions and biases guiding the prohibition discourse. This debate is the principal subject of Chapter 4. One of the key arguments is that AWS might in fact be better than humans in satisfying ethical codes and legal principles (Arkin 2010, 2017, 2018; Anderson and Waxman 2012, 2013; Schmitt 2013; McFarland 2015: 1326–1327 and 1338; Birnbacher 2016: 118–121). Bode and Huelss (2018) at the same time warned against overemphasizing 'fundamental norms' (human rights, responsibility, etc.) to the detriment of 'procedural norms' (efficiency, effectiveness, survival, obedience, etc.). All these nuances have so far prevented the prohibition discourse from becoming a 'regime of truth' (Foucault 1980: 131).

Structural power: Asymmetries in knowledge and material production

The following section analyses power hierarchies and associated interests defining the key stakeholders in the domain of AWS. In doing so, the structural position of pro-ban actors is carefully delineated in relation to ban-resisting actors. Based on the discovered distribution of benefits, obligations, and objectives, inferences are made with regard to their structural weight.

The movement working to ban AWS is composed of (international) non-governmental organizations (I)NGOs and their various umbrellas, elements of global and regional IOs, like-minded states, epistemic communities of scientists, Nobel Peace Prize laureates and Peace Prize laureate organizations, faith leaders, hi-tech business companies, and the mass media. Multiple power heterarchies, with all of the actors in different roles and capacities, are apparently involved (CSKR n.d.). Structurally, an 'effective and efficient' relationship between non-governmental or civil society organizations and governments has been gradually set up in many liberal countries (Flora 1986; Gidron, Kramer, and Salamon 1992). The former have come to be considered even better than governments at much of what they do, including the promotion of greater public awareness of international issues (Chrétien 1995). With their functions limited mainly to consultation or observation, non-governmental entities have also penetrated the UN system of governance via either formal or informal mechanisms (Ruhlman 2015: 37–40). Regularly issuing statements and organizing side events, NGOs and think tanks with participation rights have been involved in inter-governmental deliberations on possible arms control options for AWS (Rosert and Sauer 2020: 17). Some elements of global and regional IOs have praised the agenda. In the surge of lethal autonomous robotics, the report of the UNHRC's Special Rapporteur on extrajudicial, summary, or arbitrary executions has brought human rights and ethics to the fore (Heyns 2013). Endorsing a ban on AWS, the UNSG has posed a challenge to the opposition (Guterres 2018). The Organization for Security and Co-operation in Europe Parliamentary Assembly (OSCEPA) and the European Parliament (EP) have also supported the ban proposal on the parliamentary level (EP 2018; OSCEPA 2019). In reaction to the CCW process, the statement on behalf of the EU has similarly underlined the importance of human control on life-and-death decisions (EEAS 2019). The European Commission (EC) and the proposed EU Coordinated Plan on Artificial Intelligence have also promoted a 'human-centric' approach to AI (EC 2018). Acknowledged to be an important frame of securitization (Vultee 2011), media around the world have generously favoured the case. Among those having covered the agenda are high profile outlets such as the BBC, *The Independent*, *Forbes*, *The Guardian*, *The New York Times*, *The Washington Post*, *The Telegraph*, and the Thomson Reuters Foundation (e.g., CSKR 2018a). The case has also been intertwined with science fiction and fantasy, assumed to be increasingly 'invoked by policy elites in service of arguments about the real world' (Carpenter 2016: 53).

The main venue for global deliberations, particularly intergovernmental negotiations, has become the UN's CCW in Geneva (Altmann and Sauer 2017: 132). Blocked by some treaty members, mainly great powers, the CCW format has previously failed to ban APL and CM (Abramson 2017; Hynek 2018b). This is because political elites are the main actors in the global decision-making process (Basu 2004: 139). From the theoretical perspective, they are also believed to have an advantage over other actors in seeking to influence audiences and calling for the implementation of extraordinary measures (Emmers 2013: 134). Thirty states, including Algeria, Austria, Brazil, Costa Rica, Ghana, Guatemala, Jordan, Morocco, Nicaragua, Pakistan, Uganda, and Zimbabwe, have already objected to AWS (CSKR 2020). Since this does not even constitute the majority, we offer a politico-economic insight into their relative structural weight with respect to AWS regulatory politics: the global arms transfer and production system is stratified. The barriers to movement between tiers in such a system remain relatively high, whereby dominant actors concentrate on the pursuit of power and sometimes also wealth in their arms exports (Krause 1992: 98 and 206–210). Possibly for these reasons, less technologically developed states have endorsed a ban while the top world leaders in AWS development (primarily the US, China, and Russia, but also the UK, France, Germany, Sweden, Italy, South Korea, and Israel) have not done so (Haner and Garcia 2019: 332–335; CSKR 2020). It is worth elucidating, however, that a traditional circuit of 'arms transfer dependence' on the part of the recipients helps us to conceive of other less technologically developed states' opposition to the ban (Kinsella 1998: 19). Their motivation is clear: 'technologically inferior and vulnerable' states often concentrate on 'the pursuit of security' in their arms transfers (Krause 1992: 98). Therefore, opposition to the ban on AWS appears deeply embedded in the prevailing social order. Concurrently, those economically dominant actors largely belong to the Permanent Five (P5) and command the UN Security Council (UNSC), a body that has the primary responsibility of maintaining international peace and security (Ruhlman 2015: 37–38). Tailored upon the contours of business logic, their political domination recalls the destiny of cyberweapons and heralds that 'decisive political will supporting prohibition is unlikely' (Stevens 2019: 289).

Though significant for a country's economic competitiveness and export success, information technologies are among those primarily driven by commercial interests and private funds, often within transnational companies (Smith and Udis 2003: 102–103). There are indications that the commercial sector (mostly large technology companies within Western democracies) has detached from military robotics or so-called killer robots. For example, Tesla's Elon Musk and DeepMind's Demis Hassabis and Mustafa Suleyman took part in the open letter; Google pledged to not develop AI for use in weapons, having also sold the controversial Boston Dynamics; Clearpath Robotics pledged its endorsement of the preemptive ban on AWS (Altmann and Sauer 2017: 132–133; Google 2018; CSKR n.d.). While the discourse is settling, the practical commercial-military bond has become entrenched in the process of AI development in the West. Though well known at the time for its close ties to the US military, Boston Dynamics was acquired by Google for commercial purposes (Altmann and Sauer 2017: 132–133; Diakogiannis 2019). Google was also found to have had commercial ties with the US military drone programme, in particular in its contract with the Pentagon's Project Maven (Gibbs 2018; Wakabayashi and Shane 2018). While committed to its pledge, ClearPath Robotics has continued to work with its military clients (CSKR n.d.). Some of the world's largest arms producers and prominent defence contractors of the Pentagon – Lockheed Martin, Boeing, Northrop Grumman, and Raytheon – are working on technologies relevant to AWS (PAX 2019: 5–8). The idea that the private sector should detach from military robotics does not even have the same appeal in other countries. The obvious example is China. Its national strategy of 'civil-military fusion' blurs distinctions between the military and civilian domains and renders AI explicitly dual-use from the outset (Verbruggen 2019: 340). The Russian government, particularly the Ministry of Defence, has also joined forces in AI development with key national corporations and companies such as Sberbank, Yandex, Rostelecom, Rostec, Gazprom Neft, and Rosatom (Kremlin.ru 2019). The group of companies that raise particularly high concerns about AWS includes AVIC and CASC (China), IAI, Elbit and Rafael (Israel), Rostec (Russia), and STM (Turkey) (PAX 2019: 5). More generally, robotics as well as automation and complex analytics have deeply penetrated the profit-driven manufacturing industry, including tech giants such as GE, Siemens, Intel, Kuka, Bosch, NVIDIA, and Microsoft (Walker 2019). Although civilian innovation is 'no guarantee' for the development and diffusion of AWS (Verbruggen 2019: 338), both robotics and AI – technologies that pave the way for AWS – are dual-use (Altmann and Sauer 2017: 124).

Often tightly interlinked with the previous cluster of actors, epistemic communities have also engaged in deliberations over AWS security regulation (e.g., CCW 2015). This category combines scientists, technicians, defence experts, military officials, international lawyers, and even faith leaders. They are usually motivated by knowledge-based expertise and a desire to support the public interest or improve human welfare (Cross 2013: 150–158). The recognition of their field expertise or more general professional judgement may be thought of in terms of 'authority' (Schneiker 2016: Chap. 4). Framing a particular issue or problem and contextualizing it from various perspectives, they influence both governments and non-state actors (Cross 2013: 138; Schneiker 2016: Chap. 4). They have also become crucial boosters of humanitarianism (Schneiker 2016: Chap. 4). In this sense, they feature 'a major means by which knowledge translates into power'. Especially as transnational processes and transnational global governance are evolving, epistemic communities are becoming ever more significant (Cross 2013: 138–139). What is important is that members of such groups are also driven by their particular professional self-understanding (Schneiker 2016: Chap. 4). This allows for variation in internal cohesion within epistemic communities, as well as coexisting or conflicting epistemic communities (Cross 2013: 147–148). This hypothetical set-up aptly depicts the case of AWS. Some experts have articulated support for the ban (Asaro 2012; Sharkey 2012, 2017; Garcia 2015; Klincewicz 2015) while others have pointed out either biases or weaknesses in their approach and have drawn attention to positive facets of AWS (Arkin 2010, 2018; Anderson and Waxman 2012, 2013; Schmitt 2013). There is a risk of their concerns being silenced since internal cohesion within epistemic communities is essential for exercising influence on policy outcomes (Cross 2013: 138).

The ICRC deserves special attention. This neutral INGO with a legal personality and an IO-equivalent status engages in humanitarian diplomacy and occasionally prohibition-oriented policy advocacy as the cases of APL, CM, and AWS testify to (Price 1998: 632; ICRC 2016; Hynek 2018b). Its legal personality particularly implies its 'legal status vis-à-vis states in international law' (Mathur 2011: 182). It also seeks to establish and strengthen ties with the private sector (ICRC 2018). At the same time, the ICRC has emerged as the key humanitarian actor – in particular 'humanitarian expert' (Mathur 2017: 19) – first and foremost driven by its guardianship of the Geneva Conventions or IHL (Forsythe 2005: 13; Mathur 2011: 182, 2017: 4 and 95). It also possesses legal expertise in the field of IHL and IHRL, the progressive convergence of which has also entailed the convergence of their respective 'guardian institutions', namely the ICRC and the UN (Fortin 2012). Diplomatic facilitation of the pro-ban movement by the ICRC is thus of particular importance (ICRC 2016).

Admitted to have an impact on securitization processes (Salter 2008; Emmers 2013: 134), the public also gets retracted into AWS-oriented deliberations. Generally, their stance remains fragmented (Ipsos 2019, 2021). However, public opinion is contextual and varies based on the specifics of particular circumstances and scenarios (Horowitz 2016b). The strongest opposition to AWS has so far been reported in Sweden, Turkey, South Korea, and Hungary; the strongest support for AWS – in India and Israel (Ipsos 2019, 2021). While most respondents oppose AWS in the US, the UK, Russia, and China, there are still many supporters of their use in war in each of these countries. The central concern raised against AWS is that they are immoral and unaccountable (Ipsos 2019). As a survey in the US has shown, the most common reason given in favour of AWS is 'force protection', i.e., the idea that such weapons can protect the lives of human soldiers (Carpenter 2013). We assume this line of thinking represents an overall trend as '[s] ending an army of machines to war – rather than friends and relatives – does not exact the same physical and emotional toll on a population' (Wagner 2014: 50).

Overall, while pro-ban actors might be broken down into (I)NGO advocacy, state and audience support, facilitation by (elements of) IOs, expert blessing, and commercial sector endorsement, ban-resisting actors can be broken down into almost the same categories. This signifies a divided social stance in relation to the ban or severe restrictions on AWS. In terms of the structural distribution of power and associated interests, the current state of affairs does not favour the pro-ban movement. The three key clusters of stakeholders are crucial for consideration in this regard. States, especially those belonging to the P5, epistemic communities, and the commercial sector are those that primarily exercise control over the sphere of security, knowledge production and technological know-how, finance, and material production (Strange 1994: 24–32). Those most powerful states, which are concurrently the top world leaders in AWS development, oppose the ban on AWS. They particularly do so under the umbrella of knowledge-based rationale provided by experts and often in collaboration with commercial players. Their main clients in the global arms transfer and production system are also likely to remain supportive of their stance on AWS. This testifies to a disadvantageous structural position of the movement against AWS. Interestingly, many of its participants have been involved with successful APL and CM prohibition campaigns and with regulatory efforts in the case of small arms and light weapons (Krause 2002: 256; Marks 2008; Hynek 2018b). Opposition to such humanitarian endeavours has in most cases also principally lied with 'a handful of weapons-dependent governments' (Johnson n.d.). To be more specific, roughly the same few have also opposed severe restrictions on APL, CM, small arms, and nuclear weapons (Krause 2002: 256; Marks 2008; Hynek 2018b; Bourne 2019; Johnson n.d.). The following section will show how structurally disadvantaged actors might be able to overcome gaps in politico-economic power. This is particularly through certain forms of authority 'employed as a power resource to influence transnational outcomes' (Hall 1997: 591).

Compulsory power: Hybrid moral thrust vs great power politics

This section examines efforts by ban-resisting actors to maintain their structural dominance and freedom to continue investing in AWS. Also, it analyses attempts by pro-ban actors to alter their policies. Particular attention is paid to how actors' authority is translated into strategies of direct influence to either maintain or reverse power asymmetries. Empirically, we focus on the discursive and practical confrontation between the hybrid moral thrust generated by the pro-ban movement and the opposition produced by great power interests and politics.

The hybrid nature of the pro-ban movement is clearly and comprehensively illustrated in Chapter 5. It might still be evident, for example, with one of the UNGA First Committee side events held on 18 October 2016 (CSKR 2016). It brought together a person chiefly pertaining to the CCW context, a well-known scientist, and representatives of the Campaign to Stop Killer Robots, Human Rights Watch, and International Committee for Robot Arms Control. Various sources of power have constituted this hybrid force and include, inter alia, field knowledge and expert authority, informal or formal (legal) standing, and moral or political authority of (I)NGOs, especially the ICRC, and IOs, particularly the UN and its bodies. What should be considered as the central and unifying source of authority on the part of norm entrepreneurs is 'moral authority' (Hall 1997: 591). The policy of constructive engagement, most importantly within the CCW format, also offers an immense service to political activists and NGOs (Hynek 2018b).

Harnessing this leverage, pro-ban actors deployed various mechanisms to alter the behaviour of ban-resisting actors. To stigmatize opponents of their agenda and simultaneously put pressure on them, they named and shamed such countries as Russia, China, Israel, South Korea, the UK, and the US (CSKR 2019, 2020). Tabooization of AWS is also performed via imaginaries of urgency, consequential and deontological arguments, emotional appeals, and speculative fantasies (Carpenter 2016: 53). To give an example of the latter, frames from *The Terminator* and *Battlestar Galactica* have been featured in their media coverage (Carpenter 2016: 53). Religious norms have also been called upon to assist in stigmatizing AWS (PAX 2014b). To facilitate tabooization, pro-ban actors often emphasize certain attributes while ignoring other relevant traits (Sauer and Reveraert 2018: 438). Persuasion through normative and peer pressure on multiple fronts is also in their tool-kit. They bring their agenda to various conferences and events organized by governments, (I)NGOs, IOs, and groups of individuals with various expertise (CSKR n.d.). Open letters initiated by AI and robotics research communities have been broadly circulated for further signatures both locally and globally (FLI 2015; Kerr et al. 2017; Open Letter 2017a, 2017b). Another open letter, particularly targeting the CCW, has been produced by tech companies (FLI 2017). With a majority vote of participants at one of the specialized workshops, experts have issued a statement calling for a ban on AWS (ICRAC n.d.). Professional audiences, broadly categorized, also get increasingly dragged into the process, since documents of this sort often call for their endorsement via signatures (FLI n.d.). Statistics on favourable public opinion are collected and reported in the form of opinion polls as a complementary mechanism of support stimulation (Ipsos 2019, 2021). In line with its traditional mission of urging governments to adapt IHL to changing circumstances, the ICRC has also called on states to set limits on autonomy in weapons systems (ICRC 2016). It has considered its position having also collaborated with the Campaign in issuing one of its *International Review of the Red Cross* journal editions (CSKR 2013). Elements of global and regional IOs also play a crucial role in steering national ambitions in the desired direction. The declaration by the OSCEPA and resolutions by the EP have urged their member-states to work towards a legally binding ban on AWS (EP 2018; OSCEPA 2019). Some states such as Austria, Cuba, Costa Rica, Pakistan, or Guatemala are inclined towards AWS prohibition, thereby generating peer pressure on other states, especially within IOs (CSKR 2020). Actor-oriented letters and reports have also been used to put or increase normative pressure on both commercial entities (Wareham 2018a) and individual governments (Amoroso et al. 2018; Wareham 2018b). Certain reactive guarantees on the part of both (private) economic actors and states broaden the scope of peer pressure (Google 2018; Maas H. 2018; Peters 2019). Meanwhile, Campaigners regularly undertake media outreach to draw public attention (e.g., CSKR 2018a).

However, the pro-ban movement still lacks direct leverage over ban-resisting actors, as leverage maintained by politically and economically dominant states over all other actors in the equation prevails. Hall (1997: 594–597) was correct in pointing out that moral authority operates on balance with other 'power resources', mainly state economic and political(-military) capabilities. The legal status of the OSCE in relation to participating states is non-binding (OSCE n.d.). Despite its increasing prominence, the EP has little influence over member-states in the domain of EU security and defence policy (EP n.d.). The EU's political stance is still fragmented, with Austria calling for a ban while France, Germany, and the UK remaining either sceptical towards the ban or opposing it (CSKR 2020). Within the global system of governance, the UNSC stands out as 'the most structurally unequal body' (Ruhlman 2015: 37–38). Though enjoying the right of interference in internal affairs of states with humanitarian services and a broad right of initiative, the ICRC lacks credible enforcement mechanisms, particularly outside of conflict and crisis zones (ICRC 2011). The stratified global arms transfer and production system, which only emerges in the domain of AWS, usually enables dominant producers and suppliers to manipulate arms transfers and related dependencies for political ends (Krause 1992: 99–126 and 206–210; Kinsella 1998: 19). With minor exceptions, strategic partnerships run through the Shanghai Cooperation Organization, the Gulf Cooperation Council, the North-Atlantic Treaty Organization, the OSCE, the EU, and beyond also account for certain degrees of dependence and followership, particularly between smaller and greater states (Cooper, Higgott, and Nossal 1991). As the experience with nuclear weapons has shown and as now evident with killer robots, non-allied states tend to join humanitarian disarmament or prohibition agendas (Sauer and Reveraert 2018: 442). In fact, the Non-Aligned Movement has taken an active role in lobbying for prohibitions and regulations in the issue area of AWS (CSKR 2020). States may, to various extents, be in control of the commercial sector too. For example, China's national strategy of 'civil-military fusion' leaves little freedom for big companies to conduct business independently from the state (Verbruggen 2019: 340). Another example is Russia. Most of Russia's national corporations and companies that are at the forefront of AI development are, to a considerable extent, under state ownership (Kremlin.ru 2019). Both cases testify to the model of a 'developmental state' as understood by political economists (Woo-Cumings 1999).

To warrant and sustain their controversial stance on AWS, ban-resisting actors engage in what Sauer and Reveraert (2018: 446–447) call 'stigma rejection'. They recognize the stigma but detach themselves from it. For example, both Russia and the US highlight the preservation of a meaningful human element in their AI-oriented military projects (TASS 2018; BBC 2019). The US furthermore insists that, even in the event of degraded or lost communications, tele-operated systems should not be capable of autonomously selecting and engaging targets (DoD 2012/2017). Following the same course, the UK has also underlined the significance of human direction for the application of lethal force (Country Statement 2018b). China has even called for a ban on the use of fully autonomous weapons (CSKR 2020). The language of human 'involvement', 'judgement', 'control', and 'responsibility' has generally been agreed upon and formally codified at the CCW (CCW 2018, 2019a). However, the question of norm compliance remains shrouded in ambiguity, chosen by ban-resisting actors as an effective strategy to reconcile normative pressure and preference for strategic flexibility. During CCW GEE discussions, Russia has repeatedly requested to cut down on unnecessary details in the final report, especially when there has been no consensus or sufficient evidence. The US has in turn expressed concerns at the rigid concept of 'human control' (Acheson 2018; Acheson and Pytlak 2019). China's diplomatic position on AWS can also be characterized by a degree of 'strategic ambiguity'. This is evident, for instance, in its recent position papers and its national strategy of 'civil-military fusion' (Kania 2018). At the same time, it opposes a ban on the development or production of AWS (CSKR 2020). The situation is similar with private companies in the US, Russia, China, Israel, Europe, and beyond. While explicit about how human control is ensured, some continue working on technologies relevant to AWS. Quite a few do so even without clear, at least clearly declared, policies concerning the question of human control (PAX 2019: 5–8). On top of that, ban-resisting actors have simultaneously resorted to what might be termed 'counter-stigmatization' (Sauer and Reveraert 2018: 446–447). Often backed by knowledge-based expertise, they have tried to shift attention to the positive value of AWS (Country Statement 2018a, 2018b, 2019). Robert Work, former US Deputy Secretary of Defense, once argued: 'AI will make weapons more discriminant and better, less likely to violate the laws of war, less likely to kill civilians, less likely to cause collateral damage' (Fryer-Biggs 2019). A clear indication of expert support is, for example, that the Berlin Statement was not endorsed unanimously at the experts' workshop in Berlin (ICRAC n.d.). Referring to computer scientist Ronald C. Arkin, Lucas (2014) also assumed that autonomous robotic technology might 'render war itself, and the conduct of armed hostilities, less destructive, risky, and indiscriminate'. Many other frequently raised arguments against the ban are presented in Chapter 4.

As demonstrated, sheer politico-economic power is not enough: gaps in this realm can to some extent be overcome by other structures of power (discursive tools, normative pressure, moral authority, etc.). Through the latter, norm entrepreneurs are able to generate normative 'hooks' in the configuration of power, and such 'hooks' are emerging to frame AWS. Manifestations of self-restraint on the part of ban-resisting actors testify to this. However, despite the broad reach and thriving appeal of the pro-ban movement, their discourses and practices cannot curb ban-resisting actors at this moment. Politically and economically dominant states still possess credible leverage grips to mobilize opposition to the stigmatization and prohibition of AWS.

Institutional power: Diplomatic impasse and absence of legal encapsulation

In this part, it is shown how the aforementioned dynamics eventually play out in the production of *legal commitments*. This section serves to demonstrate that institutional power is largely, yet not exclusively, a product of other power relations. Conceptually, we offer insight into the mediatory power of involved institutions and the nature of indirect interactions among the key stakeholders. Decisional rules, lines of involvement and responsibility, as well as configurations of path-dependence are taken into consideration. In practical terms, the focus is on how pro-ban actors can influence or control other, especially ban-resisting, actors via institutional loci.

The CCW has firmly established itself as a humanitarian law instrument and an international forum for addressing the problematic of weapons, including AWS regulation (UNODA 2014). The practice of successfully adopted protocols on BLW, explosive remnants of war, non-detectable fragments, and other devices has set procedural and path-dependence characteristics of this negotiation format. Such prohibitive practices have a standard-setting function going far beyond a particular treaty (Rappert 2008; Acheson and Fihn 2013). Having broadened their scope to address AWS, CCW discussions were originally held in the format of informal expert meetings but gained a 'formal status' by moving to the GGE format (Rosert and Sauer 2020: 17). Over a hundred nations are state parties to the CCW, including all five permanent members of the UNSC (CSKR 2018b). The UN's CCW framework has brought together its state parties and other states, UN agencies, (regional) IOs, the ICRC, and other registered (I)NGOs, including the Campaign to Stop Killer Robots, to address AWS (UNODA 2014; CSKR 2018b). Campaigners acknowledge that it has been a 20-year practice of meaningful (I)NGO participation in such sessions (CSKR n.d.). The history of the ICRC and the UN working together is even more deeply rooted (Fortin 2012). To clarify the position by the ICRC: it has underlined the importance of human responsibility over decisions to kill, thereby granting diplomatic support to the pro-ban movement within the UN (ICRC 2016). Informal meetings between civil society and diplomats have additionally been arranged under UN auspices (CSKR n.d.). A close relationship between the UNHRC and the CCW is admitted in view of the linkage between IHRL and IHL (Bieri and Dickow 2014: 3; CCW 2015: 19). Experts, academics, researchers, and analysts have also gained a firm seat in the CCW negotiation format as recent reports also testify to (e.g., CCW 2019a). Scientists, academics, and political, military, legal, economic, and technical specialists have also appeared among governmental representatives along with ministers, secretaries, ambassadors, counsellors, and attachés (e.g. CCW 2019b).

Deeply embedded within this multi-voice institutional composition, the issue of AWS has not become institutionalized via a working international security regime. Progress in the CCW has largely been hampered by its tradition of consensus voting and lowest common denominator outcomes. Besides that, the GGE's mandate has generally been weak. Its principal goal has always been to discuss the matter and report to the CCW. The GGE has never been authorized to arrange formal negotiations for a new CCW protocol, even though it has recently been tasked to consider 'aspects' of a prospective legal framework on AWS in 2020 and 2021 (Rosert and Sauer 2020: 17). There also lack enforcement mechanisms in the CCW format (Abramson 2017).

As a result, there is no legal instrument to govern the emerging field of AWS (as of 1 September 2021). The current GGE process might be characterized as 'going slow and aiming low' (Rosert and Sauer 2020: 17). In line with the cases of APL and CM (Hynek 2018b), these are mainly dominant powers that have blocked the ban proposal (CSKR 2020). As R&D and governmental spending accelerate in the field of AWS (Haner and Garcia 2019), little progress and even a degree of regress can be observed in the GGE. The diplomatic language of a few countries, including the US, Russia, and Israel, is firmly fixed on the 'benefits' of AWS (Acheson 2018: 1). Some states criticized the GGE process for the lack of ambition and urgency as early as in 2017 (Acheson 2017: 4). The US, Russia, and Israel have even tried to remove explicit references to 'human control' – the central issue in the AWS debate since 2014 – from the GGE's latest reports (Acheson 2018: 7–8; Acheson and Pytlak 2019: 5–6). While the phrase still appears repeatedly in the 2018 report, there are barely any references to it in the 2019 report. On balance, the most recent report introduces a new guiding principle: '[h]uman-machine interaction, which may take various forms and be implemented at various stages of the life cycle of a weapon' (CCW 2018, 2019a). This provision may implicitly create space for the lawful use of AWS: the key would be to ensure that the system effectuates human intent when using force. More flexibility is given by the lack of a precise definition of AWS and a clear understanding of the relevant concepts and characteristics after six years of CCW discussions.

Increasingly frustrated with the CCW format, pro-ban actors have already begun thinking about moving the process to another venue (Bieri and Dickow 2014: 3). The UNGA or an alternative ad hoc forum seem to be 'the only legitimate spaces where progress is possible' (Acheson and Pytlak 2019: 2). Other successful treaties have demonstrated that the UNSC resistance is not necessarily decisive (Johnson, n.d.). For example, the Arms Trade Treaty or the Nuclear Weapon Ban Treaty clearly testify to the utility of the UNGA in building up international regimes. The UNGA might also be an option for AWS, especially now that regular issue-oriented deliberations at the UNGA First Committee on Disarmament and International Security have complemented the CCW procedure (UNGA 2018). Recalling the Ottawa (APL) and Oslo (CM) processes, there is also a real possibility for lesser powers led by patrons, typically middle powers, to bypass conventional arms control for an ad hoc regime (Hynek 2018b). However, neither of these options is favourable in the case of AWS, at least as of now. The UNGA upholds the principle of equality, and – on top of that – the five permanent member-states of the UNSC maintain the right to veto decisions concerning international peace and security (Ruhlman 2015: 37–38). With only China inclined towards a limited ban at the most, all of the five oppose the all-embracing ban on the use, development, and production of AWS (CSKR 2020). However, the pro-ban movement lacks not only influential members but also state members in general. This makes it much harder for them to generate any meaningful legal outcomes at the UNGA or an alternative ad hoc forum. Their coalition also lacks active middle powers whose leading role proved significant in previous norm-setting processes: Sweden on BLW, Canada on APL or Norway on CM (Rosert and Sauer 2020: 17–18).

To sum up, pro-ban actors yet remain institutionally powerless in the face of competing (mainly great powers') interests. Pursuing the process via an alternative ad hoc forum is also highly demanding, especially when there is little support from diplomatically influential governments. Benvenisti and Downs (2007: 595–597) would interpret the case as the post-Cold War trend of institutions working along 'functionalist lines', making it more convenient for powerful states and more challenging for weaker states to bargain.

Discussion and a way forward

The chapter sheds light on the complex workings of power around the security issue of AWS. We drew attention to the legal backdrop and political advancements that jointly point towards international security regulation of AWS and discerned that the legal underbrush applicable for managing AWS is yet restricted. This is because it principally features general international laws of war, in particular humanitarian principles, and the piecemeal emerging framework regulating cyberweapons. We then demonstrated how, given these limitations, the issue-oriented prohibitionary agenda has deeply penetrated the political scene cutting across global, regional, and local levels. The opening section concludes that, despite the success of 'severe politicization', AWS have not become truly 'securitized': the issue has neither moved beyond standard political procedures nor culminated in the adoption of issue-oriented, extraordinary measures. The power-analytical approach is then utilized and refined, with the nexus between ethics and law interpreted, to address the case. It serves to reflect on the dynamics and conditions influencing the prospects of outlawing AWS, as well as the ethical *and* legal intensity of the emerging regulatory framework. Attempts at establishing a global prohibition regime on AWS and the existing resistance to this are brought into sharp focus, and the main findings are graphically synthesized (Figure 6.2).

This analysis begins with the notion of *productive power* and dissects discourses and counter-discourses related to the stigmatization of AWS. It shows how both deontological and consequentialist arguments are used to frame AWS as immoral, unethical, illegal, and inhumane military instruments, and even as a threat to humankind. The imagery of killer robots is additionally utilized to build a horrifying, and withal graphic, representation of AWS. The standard of civilization is also recalled by pro-ban actors to assist in this framing. Not only does it help them to portray AWS as uncivilized, but political subjects are also differentiated on this basis into the ingroup of civilized actors and the outgroup of barbarians. We eventually argue that a *nascent stigma* is emerging in association with AWS, yet more a 'politicized' rather than actual category of weapons at the moment.

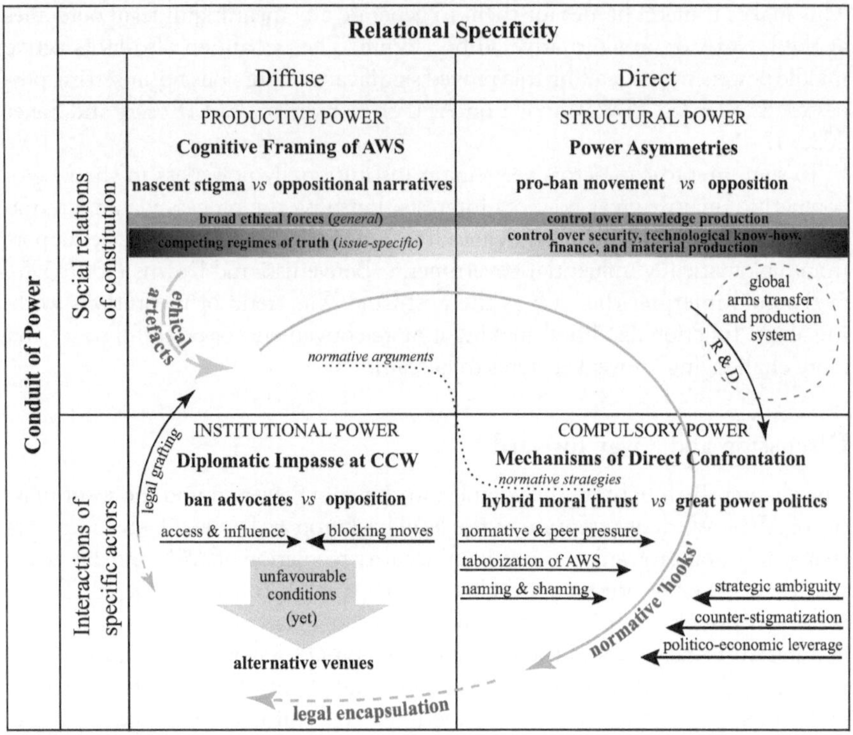

Figure 6.2 Operations of power in AWS[2]

Building upon broader ethical forces and humanitarian dispositifs, it targets the removal of human control from the use of force. The tradition of legal grafting (progressively embedding ethical codes in more and more weapons treaties) is seen as spilling over to the studied case with an incentivizing and facilitatory effect. However, we conclude that the prohibition discourse has not become a dominant regime of truth, at least as of now. Powerful oppositional narratives exist in no small part due to the lack of a meaningful material basis under the stigmatization of killer robots. They feature positive strategic-military implications of AWS and reveal the flaws of the prohibition discourse. Since ethical forces and humanitarian dispositifs are an important general condition for issue-specific discourses, competing narratives maintain a positive ethical and legal profile of AWS. Ethical artefacts that emerge in such debates form the basis for subsequent legal arguments and, prospectively, new legal principles and norms. Acceleration and increasing investments in AWS R&D, embedded in the global structure of arms production and transfers, set the whole process in motion.

Marginally touched upon in the previous section, *structural power* is our next analytical focus. Through its lens, we demonstrate how the structural position of pro-ban actors is disadvantaged in the prevailing systemic distribution of power and associated interests. To still emphasize a very high calibre of their composition, heterarchies of power underpinning the pro-ban movement are depicted. It becomes clear, however, that both pro-ban and ban-resisting actors enjoy state and public support, facilitation by (elements of) IOs, expert blessing, and commercial sector endorsement. The dynamics of arms production and trade are found to have been particularly influential in shaping the profile of ban resistance. They have affected political decision-making on the part of weapons-dependent governments, including dominant producers and suppliers as well as arms recipients, and spotlighted the role of the commercial sector. It is highlighted that economically dominant actors largely correspond to politically dominant actors in the UN system of governance (the P5). The key finding of this analysis is that a divided social stance in relation to AWS has not favoured the pro-ban movement. While the latter commands knowledge production on par with ban-resisting actors, it cedes control over the sphere of security, technological know-how, finance, and material production to them. The intensity of the gradient shade, which is even for the terrain of knowledge and uneven for the other ones, graphically nuances this (Figure 6.2).

Next, attention is directed towards the working of *compulsory power*. Here, we analyse how ban-resisting actors maintain their structural dominance and freedom to continue investing in AWS, and how ban advocates may steer their behaviour in the desired direction. Having analysed how normative arguments are translated into normative strategies, we found that structurally disadvantaged actors can harness their moral authority and to some extent overcome gaps in politico-economic power. The key mechanisms of direct influence utilized by the pro-ban movement include tabooization of AWS, naming and shaming, as well as persuasion via normative and peer pressure on multiple fronts. We draw a clear connection between their hybrid moral thrust and the emergence of normative 'hooks' working to constrain the freedom of AWS R&D. While civil society engagement and increasing public awareness are positive developments, their efforts have so far proved not enough to change the present course of affairs. Their discursive and practical strategies still concede to resistant great powers' counter-strategies. Leverage maintained by the UNSC in the global system of governance is unparalleled, compared to any other intergovernmental architecture. What reinforces the position of its members is political support by other, typically smaller and less technologically developed, states characterized by certain degrees of politico-economic dependence and followership. The commercial sector may also not have much freedom to manoeuvre, especially if governmental agencies work in close coordination with private companies. The obvious example is China where there is hardly any room for dissent against the central government. To warrant and sustain their controversial stance on AWS, ban-resisting actors address the emerging stigma as inapplicable to them but often resort to strategic ambiguity to maintain flexibility. Backed by knowledge-based expertise, they also engage in counter-stigmatization to keep the door open – just slightly – for lawful uses of AWS.

What follows is an inquiry into mechanisms of *institutional power*. The diplomatic impasse at the CCW is discovered. We recognize that ban advocates have become deeply embedded with the UN system of governance, including the CCW. Non-state actors – including (I)NGOs, representatives of the Campaign, the ICRC, scientists, academics, as well as political, military, legal, economic, and technical specialists – may access and influence institutional processes. What reinforces their institutional position is support by a group of (like-minded) governments, but also endorsement by certain elements of IOs such as the UNHRC, the UNSG, the EP, and the OSCEPA. However, the consensus rule within the CCW format, combined with the P5's implicit power to influence outcomes, renders pro-ban actors virtually powerless in the face of competing, mainly great powers', interests. There are alternative options to pursue the process: as other weapons treaties have shown, success might be reached at the UNGA or an ad hoc forum. However, the UNGA principle of equality and blocking moves by UNSC veto-bearing members may be detrimental to further progress within the UN. The lack of influential governments and sufficient mass of aligned states on the side of the pro-ban movement keeps conditions unfavourable for bypassing conventional arms control fora at this moment. Restrictions adopted athwart the opposition by dominant institutional players – largely corresponding to dominant producers of AWS – may still appear to be ineffective. This is also due to the lack of direct leverage potentially available to influence or control such non-signatories outside of institutional loci. Certain degrees of (informal) norm adherence by ban-resisting actors might even further delay progress on legal commitments, and especially a ban, in the issue area of AWS. So can the inability to agree upon a clear definition of AWS for regulatory purposes.

However, any attempt at regulating or banning weapons still hinges upon a prevailing interpretation of strategic interests. Strategic interests and political positions are not immutable; they can evolve at times bringing about outcomes previously thought to be inconceivable. An example is the Chemical Weapons Convention signed, including by the P5, after a long history of using chemical weapons in armed conflicts (World War I, Vietnam War, Iran–Iraq War, etc.). So, we cannot discard the likelihood of change in the current state of affairs on AWS, especially as a preemptive ban is just one among several options being examined. Meanwhile, the CCW can arguably continue to postpone any committal decision to another time or give states a mandate to negotiate in earnest some sort of regulation, not necessarily a ban, on AWS. Incidentally, we should not rule out at this juncture that, if CCW discussions do not produce effective results in a timely manner and a deadlock is reached by 2021, other forums or solutions may be more seriously considered to remedy the diplomatic impasse. Some still insist the CCW should be given the resources to perform its originally intended function, and propose to reform its mode of decision-making (Rosert and Sauer 2020: 20). Though a ban on AWS is unlikely to be enacted or effective (Crootof 2015: 1883–1891), an international legally binding instrument is still most actively pursued (Acheson 2017: 2). However, an international 'regime' may be complemented by additional protocols, domestic laws, as well as other informal or non-binding mechanisms (Crootof 2015: 1897–1903). Alternatively, it may be set in motion by 'interim' steps, for instance a political declaration, before legally binding measures are embraced (Acheson 2017: 2). Since there are many pathways, the dividing lines drawn in this chapter should not be seen as static. They are dynamic and contingent on changes in world politics, strategic interests, and technological developments, as well as the ability and willingness to trade-off.

Along with its empirical findings, this chapter considers how different forms of power interrelate and cross-pollinate, and how they are not alike in terms of their explanatory value. The goal is to properly grasp how the nexus between ethics and law, embedded in a complex socio-political reality, plays out in the production of new legal principles and norms. We showed how everything starts with productive power, in particular with the workings of ethical forces. We also recognized how the workings of productive power might be triggered by motions in structural power and reinforced by legal conditions already embedded in institutional power. With agents portrayed as particularly crucial for the (re)production of ethical and legal arguments, this chapter captures complex mutual relationships between productive power and all the other power configurations. It also demonstrates how enduring power asymmetries ingrained in the operation of structural power get translated into episodic manifestations of compulsory power. However, we also illustrate how gaps in structural power might be overcome by the means of compulsory power, in particular certain forms of authority translated into strategies of direct influence. The chapter culminates in showing how all these forces eventually play out to shape the contours of institutional power and the production of new legal commitments. However, the proposed trajectory connecting all types of power from ethics to legal principles and norms is not the only formula for weapons treaties. We hypothesize that, if lower strategic importance is originally attached to a weapon and a stronger stigma arises, the transposition from the structure of ethics to the legal regulatory realm may directly follow the trajectory of legal grafting (Figure 6.2). An example is the ban on BLW, whose tactical advantage was almost immediately overshadowed by unequivocal bodily harm. Since there is still more to be explored, we propose it as an avenue for further research.

Besides providing further insights into complex power interplays and ways of studying ethics and law, this study testifies to the usefulness of the power-analytical approach in general. In a departure from ontological selectiveness and agent-based explanation common in the existing literature, this approach allows for the disaggregation of a socio-political reality into a set of ontological multiplicities. At the same time, we show how the analytical findings can subsequently be synthesized into a coherent interpretation of multiple, co-existing realities and the ways in which they get intertwined and interact. Its practical application is expected to go far beyond the selected case, and the importance of future research tailoring it to other cases in the area of global governance is two-fold. First, this approach might provide valuable insights into other weapons regimes, other humanitarian regimes (e.g., refugee regimes), or security regimes (e.g., counter-terrorism regimes) outside of the realm of arms control, as well as monetary, financial, and trade regimes. Second, such investigations might help to further test the practical relevance of this approach and potentially sharpen its contours.

Concluding remarks

AWS feature a complex technological solution cultivated in civil-military synergies, and their very definition is immersed in terminological hurdles. For these reasons, AWS can hardly be equated with a perfectly identifiable weapon or device. The sci-fi imagery of killer robots, so beloved by the media, further distances AWS from the reality and calms the sense of urgency. It is also hard to isolate their negative effects for respective regulatory efforts, especially as competing narratives praise the military utility and strategic importance of AWS. There have still been enormous leaps forward on the way to stigmatizing the removal of human control from the use of force. Normative 'hooks' are emerging to frame autonomy in (lethal) weapons systems, including the current dynamics of R&D, but a taboo on AWS is hardly conceivable. As we demonstrated, ethical forces are contingent on and prone to manifold cognitive and affective manipulations. The case of AWS makes clear that they may, ironically, serve to simultaneously underpin both the prohibition discourse and competing narratives. Despite the relative advances in norm-setting, chances for the adoption of a ban or severe legal restrictions are miserably low also for other reasons. The world's acceleration and increasing investments in AWS R&D not only fuel normative debates, but also mobilize ban resistance. Efforts to curtail such projects via conventional arms control fora are likely to be hampered by the unfavourable institutional standing of their advocates. The conditions for establishing an ad hoc regime are unfavourable and deem it ineffective beforehand, as long as the world's largest producers of AWS are to be unbound by its provisions. The ever-increasing ubiquity of AI in all fields of knowledge and civil application will undoubtedly create a major verification challenge. This reveals a still significant gap between the ethical *and* legal intensity of the emerging regulatory framework. In fact, AWS do not fall outside of legal commitments altogether as international laws of war and a fragmented set of rules concerning cyberweapons are in place. However, these are disjointed and questionable in terms of their applicability to AWS. Be that as it may, it is difficult and premature to draw a definite conclusion. There is always room for new interpretations of their strategic interests by major military powers, trade-offs leading to some sort of a regulatory, possibly still prohibitory, treaty, and interim measures.

Notes

1 We judge the actors' positions not only on the basis of what they say but also on the basis of what they actually do. Most importantly, it means that the countries grouped under the banner of what we call the 'blocking coalition' do not necessarily speak publicly in favour of AWS. They have either not voted for their ban and/or engaged in the development of enabling technologies, as shown in Chapter 3.
2 Authors' own diagram, based on Barnett and Duvall (2005: 48).

References

Abramson J (2017) 'Convention on Certain Conventional Weapons at a Glance'. *Arms Control Association*, September 2017. Available at: https://www.armscontrol.org/factsheets/CCW.

Acheson R (2017) 'CCW Report', Vol. 5(6). *Reaching Critical Will*, 20 November 2017. Available at: https://www.reachingcriticalwill.org/images/documents/Disarmament-fora/ccw/2017/gge/reports/CCWR5.6.pdf.

Acheson R (2018) 'CCW Report', Vol. 6(11). *Reaching Critical Will*, 4 September 2018. Available at: https://reachingcriticalwill.org/images/documents/Disarmament-fora/ccw/2018/gge/reports/CCWR6.11.pdf.

Acheson R, Fihn B (2013) 'Preventing Collapse: The NPT and a Ban on Nuclear Weapons'. *Reaching Critical Will*, October 2013. Available at: http://www.reachingcriticalwill.org/images/documents/Publications/npt-ban.pdf.

Acheson R, Pytlak A (2019) 'CCW Report', Vol. 7(6). *Reaching Critical Will*, 21 August 2019. Available at: https://reachingcriticalwill.org/images/documents/Disarmament-fora/ccw/2019/gge/reports/CCWR7.6.pdf.

Altmann J (2009) 'Preventive Arms Control for Uninhabited Military Vehicles'. In: Capurro R, Nagenborg M (eds) *Ethics and Robotics*. Amsterdam: IOS Press.

Altmann J, Sauer F (2017) 'Autonomous Weapon Systems and Strategic Stability'. *Survival* 59(5): 117–142.

Amoroso D, Sauer F, Sharkey N, Suchman L, Tamburrini G (2018) 'Autonomy in Weapon Systems: The Military Application of Artificial Intelligence as a Litmus Test for Germany's New Foreign and Security Policy'. *Heinrich Böll Stiftung Publication Series on Democracy*, No. 49. Großbeeren: ARNOLD Group.

Anderson K, Waxman M (2012) 'Law and Ethics for Robot Soldiers'. *Policy Review* (176): 35–49.

Anderson K, Waxman M (2013) *Law and Ethics for Autonomous Weapon Systems: Why a Ban Won't Work and How the Laws of War Can*. American University Washington College of Law Research Paper 2013-11; Columbia Public Law Research Paper 13–351.

Arkin R (2010) 'The Case for Ethical Autonomy in Unmanned Systems'. *Journal of Military Ethics* 9(4): 332–341.

Arkin R (2017) 'A Roboticist's Perspective on Lethal Autonomous Weapon Systems'. In: *Perspectives on Lethal Autonomous Weapon Systems*, UNODA Occasional Papers, No. 30. United Nations Publication, New York.

Arkin R (2018) 'Lethal Autonomous Systems and the Plight of the Non-combatant'. Chapter 15. In: Kiggins R (ed) *The Political Economy of Robots*: 317–326. Switzerland: Palgrave Macmillan – International Political Economy Series.

Asaro P (2008) 'How Just Could a Robot War Be?' Part II. In: Briggle A, Waelbers K, Brey PAE (eds) *Current Issues in Computing and Philosophy*. Amsterdam: IOS Press.

Asaro P (2012) 'On Banning Autonomous Weapon Systems: Human Rights, Automation, and the Dehumanization of Lethal Decision-Making'. *International Review of the Red Cross* 94(886): 687–709.

Barnett MN, Duvall R (2005) 'Power in International Politics'. *International Organization* 59(1): 39–75.

Basu R (2004) *The United Nations: Structure and Functions of an International Organisation*. New Delhi: Sterling Publishers Private Limited.

BBC (2019) 'US Seeks to Allay Fears Over Killer Robots'. 11 March 2019. Available at: https://www.bbc.com/news/technology-47524768.

Benvenisti E, Downs GW (2007) 'The Empire's New Clothes: Political Economy and the Fragmentation of International Law'. *Stanford Law Review* 60: 595–631.

Bieri M, Dickow M (2014) 'Lethal Autonomous Weapons Systems: Future Challenges'. *CSS Analyses in Security Policy*, No. 164. Zurich: Center for Security Studies. Available at: https://css.ethz.ch/content/dam/ethz/special-interest/gess/cis/center-for-securities-studies/pdfs/CSSAnalyse164-EN.pdf.

Birnbacher D (2016) 'Are Autonomous Weapons Systems a Threat to Human Dignity?' Chapter 5. In: Nehal Bhuta et al. (eds) *Autonomous Weapons Systems: Law, Ethics, Policy*: 105–121. Cambridge: Cambridge University Press.

Bode I, Huelss H (2018) 'Autonomous Weapons Systems and Changing Norms in International Relations'. *Review of International Studies* 44(3): 393–413.

Bourne M (2012) 'Guns Don't Kill People, Cyborgs Do: A Latourian Provocation for Transformatory Arms Control and Disarmament'. *Global Change, Peace and Security* 24(1): 141–163.

Bourne M (2018) 'Powers of the Gun Process and Possibility in Global Small Arms Control'. *International Politics* 55(3–4): 441–461.

Bourne M (2019) 'Powers of the Gun: Process and Possibility in Global Small Arms Control'. Chapter 8. In: Hynek N, Ditrych O, Stritecky V (eds) *Regulating Global Security: Insights from Conventional and Unconventional Regimes*: 143–168. Cham: Palgrave Macmillan.

Brehm M (2017) *Defending the Boundary: Constraints and Requirements on the Use of Autonomous Weapons Systems under International Humanitarian and Human Rights Law*. Academy Briefing, No. 9. Geneva: Geneva Academy.

Carpenter C (2013) 'How Do Americans Feel about Fully Autonomous Weapons?' *Duck of Minerva*, 10 June 2013. Available at: https://duckofminerva.com/2013/06/how-do-americans-feel-about-fully-autonomous-weapons.html.

Carpenter C (2016) 'Rethinking the Political/-Science-/Fiction Nexus: Global Policy Making and the Campaign to Stop Killer Robots'. *American Political Science Association* 14(1): 53–69.

CCW [Convention on Certain Conventional Weapons] (2015) *Report of the 2015 Informal Meeting of Experts on Lethal Autonomous Weapons Systems*. Doc. No. CCW/MSP/2015/3. Available at: https://undocs.org/pdf?symbol=en/ccw/msp/2015/3.

CCW [Convention on Certain Conventional Weapons] (2018) *Report of the 2018 Session of the Group of Governmental Experts on Emerging Technologies in the Area of Lethal Autonomous Weapons Systems*. Doc. No. CCW/GGE.1/2018/3. Available at: https://undocs.org/en/CCW/GGE.1/2018/3.

CCW [Convention on Certain Conventional Weapons] (2019a) *Report of the 2019 Session of the Group of Governmental Experts on Emerging Technologies in the Area of Lethal Autonomous Weapons Systems*. Doc. No. CCW/GGE.1/2019/3. Available at: https://undocs.org/en/CCW/GGE.1/2019/3.

CCW [Convention on Certain Conventional Weapons] (2019b) *Provisional List of Participants Group of Governmental Experts on Emerging Technologies in the Area of Lethal Autonomous Weapons Systems*. Doc. No. CCW/GGE.1/2019/MISC.1. Available at: https://undocs.org/ccw/gge.1/2019/misc.1.

Chrétien J (1995) *Speech by Prime Minister Jean Chrétien to the National Forum on Canada's International Relations*. Toronto, 11 September 1995.

CoE [Council of Europe] (2001) *Convention on Cybercrime*. Signed 23 November 2001. Available at: https://www.europarl.europa.eu/meetdocs/2014_2019/documents/libe/dv/7_conv_budapest_/7_conv_budapest_en.pdf.

Conn A (2018) 'AI Companies, Researchers, Engineers, Scientists, Entrepreneurs, and Others Sign Pledge Promising Not to Develop Lethal Autonomous Weapons'. *Future of Life Institute*, 18 July 2018. Available at: https://futureoflife.org/2018/07/18/ai-companies-researchers-engineers-scientists-entrepreneurs-and-others-sign-pledge-promising-not-to-develop-lethal-autonomous-weapons/.

Cooper A, Higgott R, Nossal K (1991) 'Bound to Follow? Leadership and Followership in the Gulf Conflict'. *Political Science Quarterly* 106(3): 391–410.

Country Statement (2018a) *Humanitarian Benefits of Emerging Technologies in the Area of Lethal Autonomous Weapon Systems. United States of America.* Doc. No. CCW/GGE.1/2018/WP.4. Available at: https://reachingcriticalwill.org/images/documents/Disarmament-fora/ccw/2018/gge/documents/GGE.1-WP4.pdf.

Country Statement (2018b) *Human Machine Touchpoints: The United Kingdom's Perspective on Human Control Over Weapon Development and Targeting Cycles.* United Kingdom, Doc. No. CCW/GGE.2/2018/WP.1. Available at: https://reachingcriticalwill.org/images/documents/Disarmament-fora/ccw/2018/gge/documents/GGE.2-WP1.pdf.

Country Statement (2019) *Потенциальные Возможности и Ограничения Военного Применения Смертоносных Автономных Систем Вооружений* [*Potential Opportunities and Limitation of Military Uses of Lethal Autonomous Weapons Systems*]. Russian Federation, Doc. No. CCW/GGE.1/2019/WP.1. Available at: https://docs-library.unoda.org/Convention_on_Certain_Conventional_Weapons_-_Group_of_Governmental_Experts_(2019)/CCW.GGE.1.2019.WP.1_R%2BE.pdf.

Cross MKD (2013) 'Rethinking Epistemic Communities Twenty Years Later'. *Review of International Studies* 39: 137–160.

Crootof R (2015) 'The Killer Robots are Here: Legal and Policy Implications'. *Cardozo Law Review* 36: 1837–1915.

CSKR [Campaign to Stop Killer Robots] (2013) 'ICRC on New Technologies and Warfare'. 6 August 2013. Originally available at: https://www.stopkillerrobots.org/2013/08/icrc-on-new-technologies-and-warfare/. Accessed 13 November 2019.

CSKR [Campaign to Stop Killer Robots] (2016) 'Fully Autonomous Weapons and the CCW Review Conference'. Side Event Briefing, 18 October 2016. Available at: https://www.stopkillerrobots.org/wp-content/uploads/2013/03/KRC_UNGAFlyer_18Oct2016-1.pdf.

CSKR [Campaign to Stop Killer Robots] (2018a) 'Convention on Conventional Weapons Group of Governmental Experts Meeting on Lethal Autonomous Weapons Systems and Meeting of High Contracting Parties United Nations, Geneva, November 2017'. Report on Activities, 26 February 2018. Available at: https://www.stopkillerrobots.org/wp-content/uploads/2018/02/CCW_Report_Nov2017_posted.pdf.

CSKR [Campaign to Stop Killer Robots] (2018b) 'Five Years of Campaigning, CCW Continues'. 18 March 2018. Available at: https://www.stopkillerrobots.org/news/fiveyears/.

CSKR [Campaign to Stop Killer Robots] (2019) 'Minority of States Delay Effort to Ban Killer Robots'. 29 March 2019. Originally available at: https://www.stopkillerrobots.org/2019/03/minority-of-states-delay-effort-to-ban-killer-robots/. Accessed 13 November 2019.

CSKR [Campaign to Stop Killer Robots] (2020) 'Country Views on Killer Robots'. 11 March 2020. Available at: https://www.stopkillerrobots.org/wp-content/uploads/2020/03/KRC_CountryViews_11Mar2020.pdf.

CSKR [Campaign to Stop Killer Robots] (n.d.) 'A Growing Global Coalition'. Accessed 1 October 2019. Available at: https://www.stopkillerrobots.org/about/.

Dehuri S, Cho SB, Ghosh S (2011) 'Swarm Intelligence and Neural Networks'. Chapter 1. In: Cho SB et al. (eds) *Integration of Swarm Intelligence and Artificial Neural Network*: 1–21. Singapore: World Scientific.

Devlin H (2018) 'Killer Robots Will Only Exist If We Are Stupid Enough to Let Them'. *The Guardian*, 11 June 2018. Available at: https://www.theguardian.com/technology/2018/jun/11/killer-robots-will-only-exist-if-we-are-stupid-enough-to-let-them.

Diakogiannis A (2019) 'The Future of Manufacturing Technology'. *Forbes*, 6 August 2019. Available at: https://www.forbes.com/sites/columbiabusinessschool/2019/08/06/the-future-of-manufacturing-technology/#400b25e8774c.

Din AM (ed) (1987) *Arms and Artificial Intelligence: Weapon and Arms Control Applications of Advanced Computing.* Oxford: Oxford University Press.

DoD Directive (2012/2017) *Directive 3000.09: Autonomy in Weapon Systems.* US Department of Defense. Adopted 21 November 2012. Modified 8 May 2017. Available at: https://irp.fas.org/doddir/dod/d3000_09.pdf.

Dunlap C (2014) 'The Hyper-Personalization of War: Cyber, Big Data, and the Changing Face of Conflict'. *Georgetown Journal of International Affairs* 15: 108–118.

EC [European Commission] (2018) *Coordinated Plan on Artificial Intelligence.* COM (2018). Announced 7 December 2018. Available at: https://ec.europa.eu/digital-single-market/en/news/coordinated-plan-artificial-intelligence.

EEAS [European External Action Service] (2019) *EU Statement to the Group of Governmental Experts on Lethal Autonomous Weapons Systems Convention on Certain Conventional Weapons.* Geneva, August 2019. Available at: https://eeas.europa.eu/headquarters/headquarters-homepage/66584/group-governmental-experts-lethal-autonomous-weapons-systems-convention-certain-conventional_en.

Emmers R (2013) 'Securitization'. Chapter 10. In: Collins A (ed) *Contemporary Security Studies*, 3rd Edition: 131–145. Oxford: Oxford University Press.

EP [European Parliament] (2018) *European Parliament Resolution on Autonomous Weapon Systems.* Doc. No. 2018/2752(RSP). Available at: https://www.europarl.europa.eu/doceo/document/TA-8-2018-0341_EN.pdf?redirect.

EP [European Parliament] (n.d.) 'Common Security and Defence Policy'. Fact Sheet. Accessed 3 September 2019. Available at: https://www.europarl.europa.eu/ftu/pdf/en/FTU_5.1.2.pdf.

FLI [Future of Life Institute] (2015) *Autonomous Weapons: An Open Letter from AI and Robotics Researchers.* Announced 28 July 2015. Accessed 28 September 2021. Available at: https://futureoflife.org/open-letter-autonomous-weapons/.

FLI [Future of Life Institute] (2017) *An Open Letter to the United Nations Convention on Certain Conventional Weapons.* Accessed 23 September 2019. Available at: https://futureoflife.org/autonomous-weapons-open-letter-2017/.

FLI [Future of Life Institute] (n.d.) *Lethal Autonomous Weapons Pledge.* Accessed 5 October 2019. Available at: https://futureoflife.org/2018/06/05/lethal-autonomous-weapons-pledge/.

Flora P (1986) *Growth to Limits: The Western European Welfare States Since World War II. Vol. 1: Sweden, Norway, Finland, Denmark.* Berlin: Walter de Gruyter.

Forsythe DP (2005) *The Humanitarians: The International Committee of the Red Cross.* Cambridge: Cambridge University Press.

Fortin K (2012) 'Complementarity between the ICRC and the United Nations and International Humanitarian Law and International Human Rights Law, 1948–1968'. *International Review of the Red Cross* 94(888): 1433–1454.

Foucault M (1980) *Power/Knowledge: Selected Interviews and Other Writings 1972–1977.* Brighton: Harvester Press.

Fryer-Biggs Z (2019) 'Coming Soon to the Battlefield: Robots that Can Kill'. *The Center for Public Integrity*, 3 September 2019. Available at: https://publicintegrity.org/national-security/future-of-warfare/scary-fast/ai-warfare/.

Garcia D (2015) 'Killer Robots: Why the US Should Lead the Ban'. *Global Policy* 6(1): 57–63.

Garcia D (2018) 'Lethal Artificial Intelligence and Change: The Future of International Peace and Security'. *International Studies Review* 20(2): 334–341.

Gibbs S (2018) 'Google's AI Is Being Used by US Military Drone Programme'. *The Guardian*, 7 March 2018. Available at: https://www.theguardian.com/technology/2018/mar/07/google-ai-us-department-of-defense-military-drone-project-maven-tensorflow.

Gidron B, Kramer RM, Salamon LM (1992) 'Government and the Third Sector in Comparative Perspective: Allies or Adversaries?' Chapter 1. In: Gidron B, Kramer RM, Salamon LM (ed) *Government and the Third Sector: Emerging Relationship in Welfare States*: 1–30. San Francisco: Jossey-Bass Publishers.

Google (2018) 'AI at Google: Our Principles'. 7 June 2018. Available at: https://blog.google/technology/ai/ai-principles/.

Gubrud M (2014) 'Stopping Killer Robots'. *Bulletin of the Atomic Scientists* 70(1): 32–42.

Guterres A (2018) 'Remarks at "Web Summit" by United Nations Secretary-General'. Lisbon, 5 November 2018. Available at: https://www.un.org/sg/en/content/sg/speeches/2018-11-05/remarks-web-summit.

Hall RB (1997) 'Moral Authority as a Power Resource'. *Intenational Organization* 51(4): 591–622.

Haner J, Garcia D (2019) 'The Artificial Intelligence Arms Race: Trends and World Leaders in Autonomous Weapons Development'. *Global Policy* 10(3): 331–337.

Heyns C (2013) *Report of the Special Rapporteur on Extrajudicial, Summary or Arbitrary Executions*. Doc. No. A/HRC/23/47. Available at: https://www.ohchr.org/Documents/HRBodies/HRCouncil/RegularSession/Session23/A-HRC-23-47_en.pdf.

Heyns C (2016) 'Autonomous Weapons Systems: Living a Dignified Life and Dying a Dignified Death'. Chapter 1. In: Nehal Bhuta et al. (eds) *Autonomous Weapons Systems: Law, Ethics, Policy*: 3–20. Cambridge: Cambridge University Press.

Horowitz MC (2016a) 'Why Words Matter: The Real World Consequences of Defining Autonomous Weapons Systems'. *Temple International and Comparative Law Journal* 30(1): 85–98.

Horowitz MC (2016b) 'Public Opinion and the Politics of the Killer Robots Debate'. *Research and Politics* (January-March): 1–8.

HRW [Human Rights Watch] (2012) 'Losing Humanity: The Case against Killer Robots'. 19 November 2012. Available at: https://www.hrw.org/report/2012/11/19/losing-humanity/case-against-killer-robots.

HRW [Human Rights Watch] (2016) 'Making the Case: The Dangers of Killer Robots and the Need for a Preemptive Ban'. 9 December 2016. Available at: https://www.hrw.org/sites/default/files/report_pdf/arms1216_web.pdf.

HRW [Human Rights Watch] (2018) 'Heed the Call: A Moral and Legal Imperative to Ban Killer Robots'. 21 August 2018. Available at: https://www.hrw.org/report/2018/08/21/heed-call/moral-and-legal-imperative-ban-killer-robots.

HRW and IHRC [Human Rights Watch and International Human Rights Clinic] (2015) *Precedent for Preemption: The Ban on Blinding Lasers as a Model for a Killer Robots Prohibition*. Memorandum to Convention on Conventional Weapons Delegates. November 2015. Available at: https://www.hrw.org/sites/default/files/supporting_resources/robots_and_lasers_final.pdf.

Hynek N (2017) 'Regime Theory as IR Theory: Reflection on Three Waves of "Isms"'. *Central European Journal of International and Security Studies* 11(1): 11–30.

Hynek N (2018a) 'Theorizing International Security Regimes: A Power-Analytical Approach'. *International Politics* 55(3–4): 352–368.

Hynek N (2018b) 'Re-visioning Morality and Progress in the Security Domain: Insights from Humanitarian Prohibition Politics'. *International Politics* 55(3–4): 421–440.

ICRAC [International Committee for Robot Arms Control] (n.d.) 'Statements: Mission Statement, Berlin Statement'. Accessed 15 October 2019. Available at: http://icrac.net/statements/.

ICRC [International Committee of the Red Cross] (2011) 'Building Respect for the Law'. 1 May 2011. Available at: https://www.icrc.org/en/doc/what-we-do/building-respect-ihl/overview-building-respect-ihl.htm.

ICRC [International Committee of the Red Cross] (2016) 'Autonomous Weapons: Decisions to Kill and Destroy are a Human Responsibility'. 11 April 2016. Available at: https://www.icrc.org/en/document/statement-icrc-lethal-autonomous-weapons-systems.

ICRC [International Committee of the Red Cross] (2018b) 'Ethical Principles Guiding the ICRC's Partnerships with the Private Sector'. 1 March 2018. Available at: https://www.icrc.org/en/document/ethical-principles-guiding-icrc-partnerships-private-sector.

Ipsos [Institut de Publique Sondage d'Opinion Secteur] (2019) 'Six in Ten (61%) Respondents across 26 Countries Oppose the Use of Lethal Autonomous Weapons Systems'. 22 January 2019. Available at: https://www.ipsos.com/sites/default/files/ct/news/documents/2019-01/human-rights-watch-autonomous-weapons-pr-01-22-2019_0.pdf.

Ipsos [Institut de Publique Sondage d'Opinion Secteur] (2021) 'Global Survey Highlights Continued Opposition to Fully Autonomous Weapons'. 2 February 2021. Available at: https://www.ipsos.com/en-us/global-survey-highlights-continued-opposition-fully-autonomous-weapons.

Jervis R (1982) 'Security Regimes'. *International Organization* 36(2): 357–378.

Johnson AM, Axinn S (2013) 'The Morality of Autonomous Robots'. *Journal of Military Ethics* 12(2): 129–141.

Johnson R (n.d.) 'The United Nations and Disarmament Treaties'. *UN Chronicle*. Accessed 11 November 2019. Available at: https://www.un.org/en/chronicle/article/united-nations-and-disarmament-treaties.

Kania EB (2018) 'China's Strategic Ambiguity and Shifting Approach to Lethal Autonomous Weapons Systems'. *Lawfare*, 17 April 2018. Available at: https://www.lawfareblog.com/chinas-strategic-ambiguity-and-shifting-approach-lethal-autonomous-weapons-systems.

Karppi T, Böhlen M, Granata Y (2016) 'Killer Robots as Cultural Techniques'. *International Journal of Cultural Studies* 21(2): 107–123.

Kastan B (2013) 'Autonomous Weapons Systems: A Coming Legal "Singularity"?' *Journal of Law, Technology and Policy* (1): 45–82.

Keck ME, Sikkink K (1998) *Activists beyond Borders: Advocacy Networks in International Politics*. Ithaca: Cornell University Press.

Kerr I, Bengio Y, Hinton G, Sutton R, Precup D (2017) *Open Letter to the Prime Minister of Canada by Canadian AI Research Community*. Announced 2 November 2017. Available at: https://techlaw.uottawa.ca/bankillerai.

Kinsella D (1998) 'Arms Transfer Dependence and Foreign Policy Conflict'. *Journal of Peace Research* 35(1): 7–23.

Klincewicz M (2015) 'Autonomous Weapons Systems, the Frame Problem and Computer Security'. *Journal of Military Ethics* 14(2): 162–176.

Koppelman B (2019) 'How Would Future Autonomous Weapon Systems Challenge Current Governance Norms?' *The RUSI Journal* 164(5–6): 98–109.

Korać ST (2018) 'Depersonalisation of Killing: Towards A 21st Century Use of Force "Beyond Good And Evil"?' *Philosophy and Society* 29(1): 49–64.

Krause K (1992) *Arms and the State: Patterns of Military Production and Trade*. Cambridge: Cambridge University Press.

Krause K (2002) 'Multilateral Diplomacy, Norm Building, and UN Conferences: The Case of Small Arms and Light Weapons'. *Global Governance* 8(2): 247–263.

Kremlin.ru [Presidential Executive Office] (2019) 'Excerpts from Transcript of Meeting on the Development of Artificial Intelligence Technologies'. 30 May 2019. Available at: http://en.kremlin.ru/events/president/news/60630.

Krishnan A (2009) *Killer Robots: Legality and Ethicality of Autonomous Weapons*. Farnham: Ashgate Publishing.

Latour B (1999) *Pandora's Hope: Essays on the Reality of Science Studies*. Cambridge: Harvard University Press.

Lin P, Bekey G, Abney K (2008) *Autonomous Military Robotics: Risk, Ethics, and Design*. Version: 1.0.8. Prepared for US Department of Navy, Office of Naval Research. California Polytechnic State University. Available at: https://digitalcommons.calpoly.edu/cgi/viewcontent.cgi?article=1001&context=phil_fac.

Liu HY (2012) 'Categorization and Legality of Autonomous and Remote Weapons Systems'. *International Review of the Red Cross* 94(886): 627–652.

Lucas GR (2014) 'Automated Warfare'. *Stanford Law and Policy Review* 25: 317–339.

Maas H (2018) *Speech by Foreign Minister of Germany at the General Debate of the 73rd General Assembly of the United Nations*. 28 September 2018. Available at: https://newyork-un.diplo.de/un-de/20180928-maas-general-assembly/2142290.

Maas M (2018) 'Two Lessons from Nuclear Arms Control for the Responsible Governance of Military Artificial Intelligence'. Part II, Section 7. In: Coeckelbergh M, Loh J, Funk M (eds) *Envisioning Robots in Society – Power, Politics, and Public Space*. Amsterdam: IOS Press.

Marks P (2008) 'Anti-Landmine Campaigners Turn Sights on War Robots'. *New Scientist*, 28 March 2008. Available at: https://www.newscientist.com/article/dn13550-anti-landmine-campaigners-turn-sights-on-war-robots/.

Mathews RJ (2001) 'The 1980 Convention on Certain Conventional Weapons: A Useful Framework Despite Earlier Disappointments'. *International Review of the Red Cross* 83(844): 991–1012.

Mathur R (2011) 'Humanitarian Practices of Arms Control and Disarmament'. *Contemporary Security Policy* 32(1): 176–192.

Mathur R (2017) *Red Cross Interventions in Weapons Control*. Lanham: Lexington Books.

McFarland T (2015) 'Factors Shaping the Legal Implications of Increasingly Autonomous Military Systems'. *International Review of the Red Cross* 97(900): 1313–1339.

Noone GP, Noone DC (2015) 'Debate Over Autonomous Weapons Systems'. *Case Western Reserve Journal of International Law* 47(1): 25–35.

NWI [Nobel Women's Initiative] (2014) 'Nobel Peace Laureates Call for Preemptive Ban on Autonomous Weapons'. 12 May 2014. Available at: https://nobelwomensinitiative.org/nobel-peace-laureates-call-for-preemptive-ban-on-killer-robots/?ref=204.

Open Letter (2017a) *Belgian Scientists Letter on Autonomous Weapons by Belgian AI and Robotics Research Community*. Announced December 2017. Available at: https://docs.google.com/document/u/1/d/e/2PACX-1vQU8W-mpdjBqLHlA4Xgbe1BhKI4scm2UyQg3cPpylpjnOVF81OmPSE7QmzaXNDfqBeLGrNFS4ozRL8-/pub.

Open Letter (2017b) *Open Letter to the Prime Minister of Australia by Australian AI Research Community*. Announced November 2017. Available at: https://www.dropbox.com/sh/ujslcvq7224c1gw/AADADLoJV_NCbwcOsfI9n6wba?dl=0&preview=7+Nov+AI+Letter.pdf.

OSCE [Organization for Security and Co-operation in Europe] (n.d.) 'Who We Are'. Accessed 20 October 2019. Available at: https://www.osce.org/whatistheosce.

OSCEPA [Organization for Security and Co-operation in Europe Parliamentary Assembly] (2019) *Luxembourg Declaration*. Adopted July 2019. Available at: https://www.oscepa.org/documents/annual-sessions/2019-luxembourg/3882-luxembourg-declaration-eng/file.

PAX [PAX for Peace] (2014a) *Interfaith Declaration*. 1 February 2014. Available at: https://www.paxforpeace.nl/stay-informed/news/interfaith-declaration.

PAX [PAX for Peace] (2014b) 'Religious Leaders Call for a Ban on Killer Robots'. 12 November 2014. Available at: https://www.paxforpeace.nl/stay-informed/news/religious-leaders-call-for-a-ban-on-killer-robots.

PAX [PAX for Peace] (2019) 'Slippery Slope: The Arms Industry and Increasingly Autonomous Weapons'. November 2019. Available at: https://www.paxforpeace.nl/publications/all-publications/slippery-slope.

Payne K (2018) *Strategy, Evolution, and War: Apes to Artificial Intelligence*. Washington: Georgetown University Press.

Peters W (2019) *Letter to Mary Wareham, Coordinator of Campaign to Stop Killer Robots*. 1 May 2019. Available at: https://www.stopkillerrobots.org/wp-content/uploads/2019/05/NZ-Peters-Response.pdf.

Price R (1995) 'A Genealogy of the Chemical Weapons Taboo'. *International Organization* 49(1): 73–103.

Price R (1998) 'Reversing the Gun Sights: Transnational Civil Society Targets Land Mines'. *International Organization* 52(3): 613–644.

Rappert B (2008) *A Convention beyond the Convention: Stigma, Humanitarian Standards and the Oslo Process*. Land-mine Action Report, May 2008. Available at: https://brianrappert.net/images/downloads/publications/Rappert2008-A_convention_beyond.pdf.

Rosert E, Sauer F (2019) 'Prohibiting Autonomous Weapons: Put Human Dignity First'. *Global Policy* 10(3): 370–375.

Rosert E, Sauer F (2020) 'How (Not) to Stop the Killer Robots: A Comparative Analysis of Humanitarian Disarmament Campaign Strategies'. *Contemporary Security Policy*. Available at: DOI: 10.1080/13523260.2020.1771508.

Ruhlman MA (2015) *Who Participates in Global Governance? States, Bureaucracies, and NGOs in the United Nations*. New York: Routledge.

Salter MB (2008) 'Securitization and Desecuritization: A Dramaturgical Analysis of the Canadian Air Transport Security Authority'. *Journal of International Relations and Development* 11: 321–349.

Sauer F (2016) 'Stopping "Killer Robots": Why Now Is the Time to Ban Autonomous Weapons Systems'. *Arms Control Association*, October 2016. Available at: https://www.armscontrol.org/act/2016-09/features/stopping-'killer-robots'-why-now-time-ban-autonomous-weapons-systems.

Sauer T, Reveraert M (2018) 'The Potential Stigmatizing Effect of the Treaty on the Prohibition of Nuclear Weapons'. *The Nonproliferation Review* 25(5–6): 437–455.

Schmitt MN (2013) 'Autonomous Weapon Systems and International Humanitarian Law: A Reply to the Critics'. *Harvard National Security Journal Feature*: 1–37. Available at: https://harvardnsj.org/wp-content/uploads/sites/13/2013/02/Schmitt-Autonomous-Weapon-Systems-and-IHL-Final.pdf.

Schneiker A (2016) *Humanitarian NGOs, (In)Security and Identity: Epistemic Communities and Security Governance*. New York: Routledge.

Schulzke M (2017) *The Morality of Drone Warfare and the Politics of Regulation: New Security Challenges*. London: Palgrave Macmillan.

Sharkey NE (2012) 'The Evitability of Autonomous Robot Warfare'. *International Review of the Red Cross* 94(886): 787–799.

Sharkey NE (2017) 'Why Robots Should Not Be Delegated with the Decision to Kill'. *Connection Science* 29(2): 177–186.

Singer PW (2009) *Wired for War: The Robotics Revolution and Conflict in the Twenty-first Century*. New York: Penguin Books.

Smith R, Udis B (2003) 'New Challenges to Arms Export Control'. Chapter 7. In: Levine P, Smith R (eds) *The Arms Trade, Security and Conflict*: 94–110. London: Routledge.

Sparrow R (2007) 'Killer Robots'. *Journal of Applied Philosophy* 24(1): 62–77.

Stavrianakis A (2010) *Taking Aim at the Arms Trade: NGOS, Global Civil Society and the World Military Order*. New York: Zed Books.

Stevens T (2019) 'Global Code: Power and the Weak Regulation of Cyberweapons'. Chapter 13. In: Hynek N, Stritecky V, Ditrych O (eds) *Regulating Global Security: Insights from Conventional and Unconventional Regimes*: 271–296. Cham: Palgrave Macmillan.

Strange S (1994) *States and Markets*, 2nd Edition. London: Pinter.

TASS [Russian News Agency] (2018) 'Russia's Okhotnik Attack Drone to Become Prototype of Sixth Generation Fighter'. 20 July 2018. Available at: https://tass.com/defense/1014154.

Turner J (2019) *Robot Rules: Regulating Artificial Intelligence*. London: Palgrave Macmillan.

Um JS (2019) *Drones as Cyber-Physical Systems: Concepts and Applications for the Fourth Industrial Revolution*. Singapore: Springer.

UNGA [United Nations General Assembly] (2018) *First Committee Weighs Potential Risks of New Technologies as Members Exchange Views on How to Control Lethal Autonomous Weapons, Cyberattacks*. Doc. No. GA/DIS/3611. Available at: https://www.un.org/press/en/2018/gadis3611.doc.htm.

UNODA [United Nations Office for Disarmament Affairs] (2014) 'Convention on Certain Conventional Weapons'. Available at: https://unoda-web.s3-accelerate.amazonaws.com/wp-content/uploads/assets/publications/more/ccw/ccw-booklet.pdf.

Verbruggen M (2019) 'The Role of Civilian Innovation in the Development of Lethal Autonomous Weapon Systems'. *Global Policy* 10(3): 338–342.

Vultee F (2011) 'Securitization as a Media Frame: What Happens When the Media "Speak Security"'. Chapter 4. In: Balzacq T (ed) *Securitization Theory: How Security Problems Emerge and Dissolve*: 77–93. London: Routledge.

WA [Wassenaar Arrangement] (2018) *Wassenaar Arrangement on Export Controls for Conventional Arms and Dual-Use Goods and Technologies*. December 2018. Available at: https://www.wassenaar.org/app/uploads/2019/consolidated/WA-DOC-18-PUB-001-Public-Docs-Vol-II-2018-List-of-DU-Goods-and-Technologies-and-Munitions-List-Dec-18.pdf.

Wagner M (2014) 'The Dehumanization of International Humanitarian Law: Legal, Ethical, and Political Implications of Autonomous Weapon Systems'. *Vanderbilt Journal of Transnational Law* 47(5): 1371–1424.

Wakabayashi D, Shane S (2018) 'Google Will Not Renew Pentagon Contract that Upset Employees'. *The New York Times*, 1 June 2018. Available at: https://www.nytimes.com/2018/06/01/technology/google-pentagon-project-maven.html.

Walker J (2019) 'Machine Learning in Manufacturing – Present and Future Use-Cases'. *Emerj*, 23 October 2019. Available at: https://emerj.com/ai-sector-overviews/machine-learning-in-manufacturing/.

Wallach W, Allen C (2013) 'Framing Robot Arms Control'. *Ethics and Information Technology* 15: 125–135.

Walsh JI (2015) 'Political Accountability and Autonomous Weapons'. *Research and Politics* 2(4): 1–6.

Wareham M (2018a) *Letter to S. Brin, President of Alphabet Inc. and S. Pichai, Chief Executive Officer of Google*. 13 March 2018. Available at: https://www.stopkillerrobots.org/wp-content/uploads/2018/04/KRC_LtrGoogle_12March2018.pdf.

Wareham M (2018b) *Letter to Florence Parly, Minister for the Armed Forces of France*. 3 April 2018. Available at: https://www.stopkillerrobots.org/wp-content/uploads/2018/04/KRC_LtrFrance_3April2018.pdf.

Woo-Cumings M (ed) (1999) *The Developmental State*. Ithaca: Cornell University Press.

Zawieska K (2017) An Ethical Perspective on Autonomous Weapon Systems. In: *Perspectives on Lethal Autonomous Weapon Systems*, UNODA Occasional Papers, No. 30. United Nations Publication, New York.

Conclusion

This book focused on a range of varied issues and challenges related to the militarization of artificial intelligence (AI). In particular, it inquired into what is increasingly discussed by the media, scientific communities, political elites, and political activists as the AI-driven *revolution in military affairs* (AI-RMA). The discussion began with reviewing the highly complex nature of AI. We clarified that, in fact, there is no such thing as 'AI'. It was, rather, defined as an umbrella term for a broad range of advanced technologies such as machine learning, artificial neural networks, computer vision, multisensor data fusion, natural language processing, and pattern and speech recognition, which allow the mimicry of human abilities in machines. It is for this reason, at the very least, that there was still lacking a clear understanding of the dynamics of revolutionary change associated with the AI-RMA.

At the same time, there has been a burst of growth in political activism among transnational advocacy networks in opposition to the weaponization of AI. The campaign against the so-called 'killer robots' is spearheading efforts to preventively ban autonomous weapons systems (AWS). The media has actively crafted the image of killer robots featuring visuals from *The Terminator*. The agenda has deeply penetrated the political and public space and the inter-governmental mechanisms have picked up the issue for debate at the Convention on Certain Conventional Weapons (CCW) in Geneva. However, whatever progress has been achieved is modest at best, and far from a desired outcome: there is no global regime formally banning or even purposefully regulating AWS. We argue the lack of progress is at least partially connected with the unfortunate fact that the weaponization of increasingly autonomous technologies is often misunderstood and even misrepresented. Part of the problem has been that scientists seek to influence politics and many of them have become involved, through their research, in policy advocacy. This lowers the credibility of what is accepted as scientific *fact* and decreases public trust in science.

Our goal in this book was to provide the reader with a theoretically informed, empirically rich, and normatively impartial analysis of the AI-RMA. We took a broad perspective and disaggregated this revolution along multiple dimensions: technological, socio-political and institutional, historical, and spatial. We drew a comprehensive picture, offering both theoretical and practical perspectives on the militarization and weaponization of AI. Strong attention was paid to both *revolutionary* and oft-neglected *evolutionary* characteristics of the AI-RMA. In doing so, we offered an original theoretical framework based on a careful synthesis and refinement of the existing knowledge about military revolutions, technological innovations, securitization processes, and international security regimes. What is most important here is that we developed a novel conceptualization of artificially intelligent weapons which highlights the variety of evolutionary processes occurring within a particular, truly revolutionary category of weapons. The framework we propose most accurately captures the transformative potential of AI while being broadly applicable for examining military technological change and the implications thereof.

The first challenge was to grasp, conceptualize, and practically illustrate the authentic, *revolutionary* character of the AI-RMA. We showed, in a systematic and meticulous manner, that there is no single, coherent body of literature on RMA. Rather, there are different theories, approaches, and concepts and, consequently, incomplete and fragmented knowledge. We offered a way to reconcile at least some of the outstanding differences between these disjointed and piecemeal theorizations. The broader intellectual framework was borrowed from organization theory and new institutionalism. In doing so, we developed a better integrated and, in many respects, more advanced theorization of RMA. In particular, we dissected and graphically depicted the *anatomy* of radical military change. The so-called 'military affairs' were conceptualized as a *multilayered institutionalized military order*, embedded in its larger historical and inter-organizational context, as well as broader socio-cultural and politico-economic forces. Specific attention was paid to diverse organizational contexts, the multiple dimensions of interaction between the relevant actors, time lags, and, as a result, the unevenness and complexity of change. At the same time, we distinguished between revolutionary and evolutionary processes and identified suitable indicators of an *accomplished* military revolution. The latter was defined as a swift, pervasive change in strategic concepts and doctrines, the distribution of roles and responsibilities amongst military and closely related institutions, training routines, and competitive operational performance, often with social, economic, and political implications. Having established a subtle connection between the military and the rest of society in our proposed conceptualization, we simultaneously avoided the dilemma of *whether or not* military change is linked to broader changes in society.

This allowed us to bridge the gap which existed between the best available knowledge about RMA and the much more complex realities of the AI-RMA. To evaluate the applicability of our approach, we presented three case studies of states which illustrated many of the issues we discussed in relation to RMA, particularly the AI-RMA: China, Russia, and the US. Our theoretically and empirically informed conclusion was that there is indeed a revolution *underway*. Each of the three countries has crafted a new national strategy, developed a new organizational ecosystem, and introduced new ways of conducting war and providing military training in light of AI. Nevertheless, it was noted and confirmed through numerous examples that national cultural differences shape their distinct ways of thinking and distinct approaches to the realization of innovative ideas. Certain action-reaction chains were captured as well, in particular between China and the US. In conducting this analysis, we highlighted that close integration of civilian and military technology development drives much of the innovation in AI. Be that as it may, the revolution itself is *not fully accomplished*. There is still much to be done in defining doctrinal approaches and operational principles, as well as achieving the overall operational efficiency and effectiveness expected to result from the deployment of artificially intelligent military technologies. And yet, it will take time for the revolution to spread globally. It becomes increasingly clear, however, that this revolution has important social and political implications. Throughout this book, we discussed how civilian concerns may prevent, or at least hinder, certain military innovations. The fear of intelligent machines that could possibly kill people *autonomously* has provoked a coordinated response by civil society: the Campaign to Stop Killer Robots. The Campaign has *activated* the broader inter-organizational field and united international, regional, and national non-governmental organizations (NGOs), technology companies, expert communities, like-minded governments, administrative officers, and international organizations (IOs). Being a direct response to ongoing revolutionary changes and visionary R&D, the Campaign is, in principle, about *steering* the course of the AI-RMA. It has actively lobbied for a complete global ban on a certain category of weapons associated with this RMA: *fully* autonomous *lethal* weapons systems. The discussions have, as a result, moved onto an inter-governmental track and took the form of systematic negotiations at the CCW.

We went one step further still and inquired into the *evolutionary* character of the AI-RMA. This aspect has often been overlooked, or perhaps deliberately suppressed, in favour of the argument for a new military 'revolution'. What motivated us to pursue this line of research in the first place was the lack of a coherent framework from which to conceptualize revolutionary technologies themselves. We found it to be a serious obstacle to understanding the transformative potential of AI. With this in mind, we carried out a *diachronic* analysis of a long history of technological development and identified several evolutionary trends. Numerous examples were used to demonstrate that knowledge and technology have accumulated over time. This allowed us to move beyond the most prevalent, linear and chronological, historical perspectives on military technology. Instead, we offered an integrated technological model depicting what we called *composite systems*. Such systems, as we showed, are constituted by four *layers* of systems capabilities spread across five warfighting domains. Not only did the model capture the complex nature of AI, itself composed of both emerging and long-existing technologies, but it also – more importantly – accounted for their incorporation into machines with familiar physical characteristics such as drones. Our contribution thus lied in showing that the dominant, ideal typical representations of military-technological innovation ('generations', 'ages', etc.) are not so much accurate as generally believed. It equally consisted of striking a balance between evolutionary and revolutionary approaches to understanding technological and military change. This is because the model was, at the same time, presented and treated as the *technological parameter* of RMA.

Having designed the model, we shifted the focus from theory to practice, and particularly to technical aspects of 12 *composite systems*: F-35, Ripsaw M5, Sea Hunter, and Chameleon Constellation (US); S-70 Okhotnik, Pantsir-SM, Poseidon, and RB109-A Bylina (Russia); and Blowfish A2, Sharp Claw I, Marine Lizard, and Scavenger Satellites (China). Their selection was driven by our effort to demonstrate the breadth and complexity of the ongoing *militarization* of AI. Weapons systems, other military systems and dual-use systems, weapons platforms and munitions, manned and unmanned systems, air, ground, surface, outer space, and even electronic warfare systems were among our carefully selected case studies. Our original argument for their possible *compositeness* appeared to be applicable across the board. AI technologies were indeed found to represent a new *layer* of systems capabilities deposited onto the three already existing layers. One of the key messages here was that technologies do not stop being *tools* in the hands of humans, even if endowed with more and more complex functions. Going a step further, we singled out, compared, and categorized increasingly autonomous weapons systems to get a better understanding of the *weaponization* of AI. AI was identified as a major enabler of six intelligent *functions* that are being developed in next-generation weapons: (1) enhanced situational awareness; (2) decision support; (3) target acquisition; (4) interoperability; (5) recognition of human-selected targets; and (6) lethal decisions in controlled circumstances. Our final conclusion was the following: AI technologies do not signal a *clean* break with the past but bring *new capabilities* to the whole range of military systems, all deployed or activated by humans. Most importantly, we provided convincing evidence that *fully* autonomous *lethal* weapons systems, a.k.a. 'killer robots', especially those that look and behave like the Terminator, do not yet exist.

However, we showed that the lack of a meaningful material basis is not the only problem with the killer robots discourse and the Campaign to Stop Killer Robots. Throughout the book, three other major problems were identified and explained. First, we found and comprehensively demonstrated that the prohibition discourse seriously downplays the advantages of AWS. It invites fair criticism, unsurprisingly, so there has been a proper debate, which we dissected and restructured in this book. Our primary goal was to draw attention to a dichotomy that underlies the debate on AWS. In particular, we presented a thorough analysis of seven specific *dilemmas*, detailing both pro and con arguments for each: (1) the issue of whether controllability and predictability are necessary and achievable in the age of AWS; (2) the prospect of further dehumanization of killing; (3) the prospect of further depersonalization of the enemy; (4) the deployment of human forces alongside AWS; (5) wider political and strategic considerations relating to the development and proliferation of AWS; (6) the continued applicability of existing legal frameworks, mainly the laws of war, to the possible use of AWS; and finally (7) cultural dilemmas raised by AWS. In line with the general purpose of this book, the analysis included a nuanced discussion of novelty vs continuity in narratives around AWS. We arrived at the conclusion that such weapons, if ever developed, would be a mixture of good and bad, old and new elements. Only then could we seriously argue that without *careful* consideration of potential benefits and risks, a simple solution in the form of a total and preventive ban might perhaps miss the complexity of the matter in question.

But here comes the second point: there are more perplexing problems *within* the Campaign to Stop Killer Robots. Besides discussing the flawed nature of the empirically inaccurate and one-sided killer robots discourse, its cognitive construction, representation, and promotion were further problematized. We focused on the paradox of why the (seemingly) most *efficient* means do not necessarily lead to the most *effective* solution of the problem. Lacking a strong theoretical framework for how to approach and explain this contradiction, we utilized elements of securitization theory but organized them in a way that allowed us to introduce new logics related to the problem of *over*-securitization. Two concepts were introduced to theorize – and then trace empirically – two mechanisms through which over-securitization operates: *hybridization* and *grafting*. Through these theoretical lenses, we were able to understand the paradoxical effects of two separate, yet interrelated, dimensions of the securitization process, respectively: broadening the stakeholder base and generating the sense of danger and urgency. The patterns we uncovered helped us comprehend why the Campaign's highly *efficient* efforts have not yet translated into *effective* and sustainable political success. We found that its legitimate authority is increasingly diffused with more and more actors involved, which contributes to greater breadth of the movement but simultaneously results in multiple, sometimes contradictory and not necessarily reliable, flows of information. At the same time, we discovered that the Campaign and other relevant contributors have sometimes twisted and even distorted facts to create and spread the sense of imminent danger. It is quite understandable as they intend to convey the message in a simple, easy-to-understand fashion. However, this has often led them away from the desired goal, rather than closer to it, because controversial statements and especially false arguments are easier to disprove. For example, references to *The Terminator*, warnings of the threat of an arms race in military AI, or parallels drawn between AWS and weapons of mass destruction do indeed add to the sense of urgency but can be dismissed as unrealistic, imprecise, or unreliable on the basis of bare facts.

Our next and final step was to examine the resistance met by the Campaign to Stop Killer Robots. The primary underlying goal was to evaluate the prospects for arms control in the issue area of AWS. We employed the singular, yet multifaceted, concept of *power* as an analytical tool to inquire into the multiplicity of power relations between the humanitarian disarmament campaign and the established global politico-economic structures which drive R&D. Their encounters were examined at all levels: discourses, structural relations, coercive strategies, and positions within relevant security-related institutions. Making the concept of power, rather than any specific theoretical approach, central, we reached a high degree of flexibility. Furthermore, we took significant steps beyond the available literature to theorize synergies between different types of power – productive, structural, compulsory, and institutional – and traced their complex interactions in practice. This allowed us to carefully consider discursive, normative, legal, cultural, institutional, political, economic, strategic, military, and technological dynamics, as well as interstices and patterns of mutual influence between them as part of a joint investigation. Our comprehensive analysis revealed a considerable gap between the *ethical* and *legal* intensity of the emerging regulatory framework. On one hand, a nascent stigma is emerging in association with AWS. Normative 'hooks' are being set up to constrain the freedom of R&D. However, barely any progress has been made towards actually ensuring a robust international legal framework. We explained this controversy by showing that the powerful humanitarian discourse cannot, at least for now, fully compensate for the pre-established structural asymmetries and discriminatory institutional arrangements. It is the third major problem.

The findings have three important implications for understanding the general dynamics of arms control and disarmament. First, we drew attention to the potentially deleterious consequences of over-reliance on a flawed, sometimes even blatantly wrong, representation of the particular weapons in question. Throughout this book we showed that the very term 'killer robots', and especially retro-futuristic fictional images from *The Terminator*, produce an exaggerated reality that seeks to facilitate the stigmatization of AWS. However, we made clear that such a narrow, single-sided representation of artificially intelligent weapons puts aside other uses of AI. It was found and demonstrated in numerous examples that AI is primarily designed to perform *supporting* functions in military systems and networks, with only limited and very specifically defined autonomy related to the use of force reported to be under development. AI has not been weaponized at full capacity, at least for now, and is unlikely to be deployed against human beings on the battlefield in the foreseeable future. Therefore, political discourse has apparently overrun practice. What policy advocates are attempting here is not something entirely new – a similar tendency was observed earlier, for example, with blinding laser weapons (BLW). It was an excellent illustration of how successful political advocacy can be in forging a *preventive* ban on a whole category of conventional weapons still 'in the prototype phase of development' (Akerson 2013: 70). The Protocol on Blinding Laser Weapons (1995) targeted a narrow enough category of weapons with a well-defined harmful effect: 'weapons designed to cause permanent blindness' (Sivakumaran 2012: 399). However, the same cannot be said about AWS. AWS lie at the crossroads between weapons, ways in which weapons are deployed (i.e., shifted to a fully autonomous mode), and perhaps even warriors (i.e., the prospect of human soldiers being replaced, at least partially, with sufficiently autonomous weapons). Rather unsurprisingly, the risks and benefits remain unclear. We provided convincing evidence that in a case like this, engaging in a proactive and imaginative, rather than reactionary and fact-based, discourse may have adverse effects on the future of arms control, and this leads us directly to our second point.

As far as we were able to observe, there has eventually been a split between two different, mutually insulated tracks: the discourse of disarmament advocates and the much more complex realities of military R&D. China, Russia, and the

US, as illustrated in the book, have invested heavily in military systems of a different quality from those imagined by the Campaign to Stop Killer Robots. AI is being utilized to support rather than replace human decision makers. Its particular functional role is to assist humans, sometimes to a considerable degree, in tactical navigation, collection and analysis of critical data, human-machine and machine-machine communication and coordination, as well as target detection, recognition, and acquisition, etc. Only in very rare circumstances, and typically in a *controlled* environment, will AI itself decide upon and execute the attack. Such weapons, however, are either intended to have non-lethal effects (e.g., Pantsir-SM, or RB109-A Bylina, as discussed in Chapter 3), or are incapable of self-activation and will be deployed tactically in areas of active combat where their military utility increases considerably (e.g., Kalashnikov's neural network based combat module, as cited in Chapter 4). Therefore, we put forth the following problem: increasingly vociferous arguments against the so-called 'killer robots', resting on a flawed representation of technology, cannot successfully withstand the test of criticism and may very soon start losing their appeal and relevance. However, there was no question of establishing the importance of the Campaign. Our intention was only to show that there will not be any major progress in arms control until these two tracks meet, that is until the humanitarian disarmament discourse accurately reflects and reacts more adequately to the realities of R&D.

The third major lesson to be drawn from the present analysis concerns the role of experts that actively and willingly contribute to arms control policymaking. Discussing the ultimately irresolvable contradiction between one's moral conviction and rational reasoning, Max Weber argued in his famous lecture 'Politics as a Vocation', delivered on 28 January 1919, that '[p]olitics is made with the head, not … the soul'. Peter M. Haas (1992) clarified that decision makers increasingly find themselves in situations of uncertainty and the role of knowledge-based experts, or so-called epistemic communities, is to 'ameliorate the uncertainties'. What sets the latter apart from, for example, social movements or interest groups, according to his interpretation, are their 'claims to knowledge, supported by tests of validity'. However, we observed the opposite trend in the studied case: the distinction between science and political advocacy has become ever more blurred, with experts relying heavily on emotional appeal and much less on factual evidence. Their direct engagement in politics and advocacy activity, with statements not always backed by unbiased scientific fact, may erode public trust in science and expertise. From this position, it may become increasingly difficult for them to promote any good cause. The case is not unique: what we are increasingly observing over the last two or three decades is the birth of scientific and academic policy advocacy (Hynek and Chandler 2013). It should also not be forgotten that there has been no – and little attempt at achieving – a mutually satisfying and truly rational expert consensus on AWS.

Besides the more general lessons, the book also identified three major themes that will dominate any future discussion on AWS. First of all, there is still a lot of work to be done to properly delimit the category of weapons that is particularly problematic. We explained there is much difference between *militarized* AI,

AI-*assisted* weapons, and *weaponized* AI. Examples were given to demonstrate that AI can be tailored to specific military uses, not necessarily involving weapons (e.g., Sea Hunter, as discussed in Chapter 3). AI can alternatively be incorporated into weapons systems and execute many different kinds of functions, other than autonomous firing (e.g., S-70 Okhotnik, as illustrated in Chapter 3). In rare cases, as already mentioned above, it can take over the final decision to open fire or attack other types of targets, for example, in electronic space (e.g., Pantsir-SM and RB109-A Bylina, respectively, as cited in Chapter 3). However, truly autonomous *lethal* AI does not yet exist and, in fact, is all too far removed from the reality of today's R&D. Some attempts in this direction have been made but they are rather modest and limited only to the development of a *switchable mode* which would allow a balance between human control and system autonomy depending on the battlefield situation (e.g., Kalashnikov's neural network based combat module, as referred to in Chapter 4). The very possibility of such a mode will be the key stumbling block on the way towards some sort of a regulatory, or possibly still prohibitory, treaty, the reason being that it will be virtually impossible to determine whether any particular decision to kill is made with the autonomous mode being on or off at the present moment.

Another pressing issue is the format of future negotiations. Pro-ban actors, as comprehensively discussed in this book, have been increasingly frustrated with the CCW. Some of them have already come out with another preferred option: to move beyond the traditional format of negotiations. We discussed two possibilities which derive from previous experience. The first is to adopt the ban with exemptions and based on the consent of *contracting parties* at the United Nations General Assembly (UNGA). Indeed, great power resistance is not necessarily decisive, as demonstrated by the Arms Trade Treaty and the Nuclear Weapon Ban Treaty. The second possible option is to pursue an ad hoc process, as perfectly illustrated by the two specific weapons bans on anti-personnel landmines (APL) and cluster munitions (CM), better known as the Ottawa and Oslo Conventions. However, we argued that neither of these options is now particularly promising for the Campaign to Stop Killer Robots. Relatively few states have so far supported the all-embracing ban on the use, development, and production of AWS. It will, therefore, be difficult to generate any meaningful legal outcomes at the UNGA. The pro-ban coalition also lacks active middle powers whose leading role proved significant in previous, especially ad hoc, norm-setting processes: Sweden on BLW, Canada on APL, or Norway on CM. Even if eventually adopted by a group of like-minded states, the ban on AWS will likely suffer a fate that resembles that of the Nuclear Weapon Ban Treaty. It will certainly be a step in the direction of establishing responsible state practice but will be of very limited practical relevance. With only China inclined towards a limited ban on AWS at the most, the other top world leaders in the development of relevant technologies seek to prevent, or at least to postpone, any severe restrictions. R&D will, therefore, remain unbound until they sign the ban treaty. However, the emerging stigma can possibly push states towards some sort of a legally binding regulatory treaty or, perhaps for starters, a legally non-binding political declaration.

What will give a clear focus to present and future discussions is that which Hutter and Lloyd-Bostock (1990) called 'the power of accidents'. Here comes the paradox. The Campaign calls for the *preventive* prohibition of AWS. However, the particular weapons in question are so difficult to define and so obviously complex that repeated reports upon actual accidents and violations of international humanitarian law might be the only key to an arms control regime. Such accidents are inevitable, especially as states do not possess the monopoly of technologies and know-how, which are increasingly *dual-use* as highlighted consistently throughout this book. Two major questions will subsequently drive decisions regarding the legality of AWS. That is whether such weapons can be safely deployed alongside fellow human soldiers in view of the possibility and fear of friendly fire incidents; and whether it is possible to identify the bearer of legal and political responsibility in case something goes wrong and an innocent human being – whether soldier or civilian – is killed by an 'algorithm'.

To conclude, we wish to reflect on the broader implications of the arguments presented in this book and possible avenues for further research. The first and most important finding was that both evolutionary and revolutionary changes constitute the AI-RMA. We introduced the concept of *composite systems* – and properly visualized it – to capture the former, an aspect that is neglected at best or ignored at worst in the current literature. The presented model will have a lasting relevance principally because its parameters are not case dependent. Composite systems, as we argued, are constituted by *layers* of more and more advanced technological capabilities. The newly added layers are often revolutionary in terms of outcomes but grafted onto each other through evolutionary processes. That said, it is not a specific concept relegated solely and exclusively to the AI-RMA, but has a more general applicability. Yet further research is required to prove unequivocally that the concept and model are indeed of general importance for understanding military-technological innovation.

In the absence of coherent theoretical and practical knowledge about RMA, we also developed an integrated yet open intellectual framework for thoroughly capturing the revolutionary character of the AI-RMA. Here we distinguished between revolutionary changes in the strict sense and broader evolutionary changes within and across different levels of military organization. The framework draws on the richness of the existing knowledge about the nature and history of RMA. Therefore, it is also not case-dependent. Further research efforts are necessary to determine the general applicability of our approach to the analysis of both radical and incremental transformations in military affairs at all possible levels. On a more general note, our findings on the complex dynamics of technological innovation and military change could be used to better understand the 'evolution'/'revolution' dichotomy. This knowledge would be applicable both to the field of international security studies and more general social scientific theorizations.

Furthermore, this book shed additional light on processes related to the formation of international security regimes with a two-fold contribution. First, we contributed to a better understanding of the *continuum* of securitization. We proved that securitization efforts are not a matter of dichotomy between success and failure, but rather of degree and quality. In particular, we focused on the problem of *over*-securitization and conceptualized the two underlying mechanisms, as also explained above. This problem has heretofore not received the attention it deserves. We took the first step towards this aim, and further research is encouraged to understand why over-securitization occurs and how it plays out in practice. Second, we offered an original approach to regime theorization, striking a balance between the three waves of theorizing regimes within the discipline of IR. It became possible with the multifaceted concept of *power*, applied and further refined to observe the reality as it was, in all its complex manifestations. The *power-analytical approach*, as we called it, allowed us to avoid ontological selectiveness, a flaw common in the existing literature. Rather than focusing on particular actors and their actions – be they humanitarian activists or great powers – we observed the entire spectrum and seriously considered a wide range of structural forces, enabling, shaping, or constraining the course of events. It means we developed an approach that may serve as a benchmark for studying other security regimes. In doing so we also offered, as one of our key contributions, a novel theorization of how different types of power interrelate and cross-pollinate. We encourage more research in this direction too.

Our intention behind writing this book was both to provide a robust and innovative social scientific theorization in this issue area and to facilitate a thoughtful dialogue about the true nature of the AI-RMA and AWS. We do not expect everyone to agree precisely with our take on the subject matter and related conclusions but expect more open and constructive academic and policy debates to follow. If the dialogue – both academic discourse and policy making discussions – fails, any effort being made to handle the ongoing military revolution will not be very effective, as observed in previous years and comprehensively discussed in this book.

Bibliography

Akerson D (2013) 'The Illegality of Offensive Lethal Autonomy'. Chapter 3. In: Saxon D (ed) *International Humanitarian Law and the Changing Technology of War.* Boston: Martinus Nijhoff Publishers.

Haas PM (1992) 'Introduction: Epistemic Communities and International Policy Coordination, International Organization'. *Knowledge, Power, and International Policy Coordination* 46(1): 1–35.

Hutter BM, Lloyd-Bostock S (1990) 'The Power of Accidents: The Social and Psychological Impact of Accidents and the Enforcement of Safety Regulations'. *The British Journal of Criminology* 30(4): 409–422.

Hynek N, Chandler D (2013) 'No Emancipatory Alternative, No Critical Security Studies'. *Critical Studies on Security* 1(1): 46–63.

Sivakumaran S (2012) *The Law of Non-International Armed Conflict.* Oxford: Oxford University Press.

Index

Note: *Italic* page numbers refer to figures and page numbers followed by "n" denote endnotes.